ENDOCRINOLOGY RESEARCH AND CLINICAL DEVELOPMENTS

USES OF ELECTRICAL STIMULATION FOR DIGESTIVE AND ENDOCRINE SURGEONS

ENDOCRINOLOGY RESEARCH AND CLINICAL DEVELOPMENTS

Additional books and e-books in this series can be found on Nova's website under the Series tab.

NEW DEVELOPMENTS IN MEDICAL RESEARCH

Additional books and e-books in this series can be found on Nova's website under the Series tab.

ENDOCRINOLOGY RESEARCH AND CLINICAL DEVELOPMENTS,
NEW DEVELOPMENTS IN MEDICAL RESEARCH

USES OF ELECTRICAL STIMULATION FOR DIGESTIVE AND ENDOCRINE SURGEONS

JAIME RUIZ-TOVAR

EDITOR

Copyright © 2019 by Nova Science Publishers, Inc.

All rights reserved. No part of this book may be reproduced, stored in a retrieval system or transmitted in any form or by any means: electronic, electrostatic, magnetic, tape, mechanical photocopying, recording or otherwise without the written permission of the Publisher.

We have partnered with Copyright Clearance Center to make it easy for you to obtain permissions to reuse content from this publication. Simply navigate to this publication's page on Nova's website and locate the "Get Permission" button below the title description. This button is linked directly to the title's permission page on copyright.com. Alternatively, you can visit copyright.com and search by title, ISBN, or ISSN.

For further questions about using the service on copyright.com, please contact:
Copyright Clearance Center
Phone: +1-(978) 750-8400　　　　Fax: +1-(978) 750-4470　　　　E-mail: info@copyright.com.

NOTICE TO THE READER

The Publisher has taken reasonable care in the preparation of this book, but makes no expressed or implied warranty of any kind and assumes no responsibility for any errors or omissions. No liability is assumed for incidental or consequential damages in connection with or arising out of information contained in this book. The Publisher shall not be liable for any special, consequential, or exemplary damages resulting, in whole or in part, from the readers' use of, or reliance upon, this material. Any parts of this book based on government reports are so indicated and copyright is claimed for those parts to the extent applicable to compilations of such works.

Independent verification should be sought for any data, advice or recommendations contained in this book. In addition, no responsibility is assumed by the Publisher for any injury and/or damage to persons or property arising from any methods, products, instructions, ideas or otherwise contained in this publication.

This publication is designed to provide accurate and authoritative information with regard to the subject matter covered herein. It is sold with the clear understanding that the Publisher is not engaged in rendering legal or any other professional services. If legal or any other expert assistance is required, the services of a competent person should be sought. FROM A DECLARATION OF PARTICIPANTS JOINTLY ADOPTED BY A COMMITTEE OF THE AMERICAN BAR ASSOCIATION AND A COMMITTEE OF PUBLISHERS.

Additional color graphics may be available in the e-book version of this book.

Library of Congress Cataloging-in-Publication Data

ISBN: 978-1-53615-036-0

Published by Nova Science Publishers, Inc. † New York

To my family:

To Carolina, my wife and eternal support, friend and colleague

To Jaime and Mateo, my sons and source of happiness

To my parents, Gregorio and Celia, without their lessons of life, it could not be possible for me to be writing this book now.

CONTENTS

Preface		ix
Chapter 1	Mechanism of Action of Peripherical Electrical Stimulation (TENS and PENS) *Pablo Priego*	1
Chapter 2	A Percutaneous Electrical Neurostimulation of the Posterior Tibial Nerve for Fecal Incontinence *Maria del Mar Aguilar-Martínez, Luis Sánchez-Guillén and Antonio Arroyo*	17
Chapter 3	Percutaneous Electrical Neurostimulation of Dermatome T6 for the Treatment of Obesity *Jaime Ruiz-Tovar and Carolina Llavero*	31
Chapter 4	Percutaneous Electrical Neurostimulation of Dermatome T7 for the Treatment of Diabetes Mellitus *Jaime Ruiz-Tovar and Carolina Llavero*	47

Chapter 5	Postoperative Percutaneous Abdominal Electrical Stimulation (PPAES) for Ileus in Patients Undergoing Colorectal Resection *Pedro Moya, Manuel Ferrer-Márquez, Francisco Rubio-Gil, Rafael Calpena and Angel Reina*	55
Chapter 6	Electrical Stimulation for the Management of Postoperative Pain *Andrés García Marín and Mercedes Pérez López*	69
Chapter 7	Electrical Stimulation for Improving Physical Fitness Pre- and Postoperatively after Abdominal Surgery *Artur Marc Hernández*	79
Chapter 8	Intraoperative Neuromonitoring of the Recurrent Laryngeal Nerve *Manuel Durán Poveda, Leire Zarain Obrador, Alejandro García Muñoz-Najar, Jaime Ruiz-Tovar and Gianlorenzo Dionigi*	125
About the Editor		145
Index		147

PREFACE

In the last decades, medical practice is focused of minimally invasive approaches, trying to reduce the impact of any diagnostic or therapeutic techniques on the patient. These methods are mainly focused on anatomical basis, trying to reduce surgical or endoscopic incisions, punctures or just damages.

However, in the last years, the use of electrical devices has been widely developed for different aims. In these cases, the theoretical background for their development was the physiology of electrical conduction throughout the nerves. Hereby, different artificial reflexes have been developed with medical aims, trying to stimulate several organs, such as pancreas, stomach or small bowel. On the other hand, the nervous conduction has been also used for pain release and for muscular hypertrophy and improvement of physical fitness. Finally, with just diagnostic aims, the neuromonitorization of the recurrent laryngeal nerve is a useful tool to reduce the risk of intraoperative nervous damage during thyroid or any other neck surgery.

I hope that this book, based on the current evidence of electrical stimulation methods can be helpful in the clinical practice. However, we must keep always in mine, that medical investigation obtains new data, drugs and approaches every day, so that current evidence can be outdated in the following decade, requiring future updates.

Finally, I want to thank to all the contributing authors, all of them friends and experts in their corresponding fields, their availability, time and efforts to write all the chapters.

Jaime Ruiz-Tovar, MD, PhD
Editor

In: Uses of Electrical Stimulation...
Editor: Jaime Ruiz-Tovar

ISBN: 978-1-53615-036-0
© 2019 Nova Science Publishers, Inc.

Chapter 1

MECHANISM OF ACTION OF PERIPHERICAL ELECTRICAL STIMULATION (TENS AND PENS)

Pablo Priego[*]
Department of General Surgery,
Division of Esophagogastric and Bariatric Surgery,
Ramón y Cajal University Hospital, Madrid, Spain

ABSTRACT

Electrotherapy is the science for treatment of injuries and diseases by means of electricity. Basically, electrotherapy is based on the following types: transcutaneous electrical nerve stimulation (TENS) and percutaneous electrical nerve stimulation (PENS). Both systems produce their analgesic effect by activation of afferents of deep tissues by stimulation of primary Aβ large diameter afferent fibers. The physiological mechanisms of action of TENS at low and high frequency are different, although both occur at peripheral, spinal and supraspinal level, and are

---
[*] Corresponding Author Email: papriego@hotmail.com.

based primarily on the activation of different opioid receptors. Several studies in animals and clinical investigations have been conducted, in order to elucidate the physiological effects produced in the body when electrical nerve stimulation is applied.

Keywords: electrotherapy, transcutaneous electrical nerve stimulation (TENS), percutaneous electrical nerve stimulation (PENS), mechanism, action

INTRODUCTION

Electrotherapy is the science for treatment of injuries and diseases by means of electricity.

Application of electricity for pain treatment dates back to thousands of years before Christ (BC). It is known that Ancient Egyptians applied the Nile catfish for treatment of painful conditions, as it has been shown in some tomb paintings as early as 2500 BC [1]. In the same way, there are also allusions in the Greek and Roman physicians, who prescribed the direct contact with the ray fish (black torpedo fish) for treatment of patients with gout, arthritis or headaches (Figure 1). In the 18th century, the invention of electrical devices helped to popularize electrotherapy and pain treatment by natural electricity was replaced. The 19th century was the "golden age" of electrotherapy. It was applied for countless dental, neurological, psychiatric and gynecological disturbances. However, the increasing use of effective pharmacological treatments at the beginning of the 20th century, and the lack of scientific basis of natural electricity treatment, decreased the use of electrotherapy [2]. In the second half of the 20th century, and based on animal experiments and clinical investigations, the scientific bases of electrotherapy were increasingly elucidated.

Two principal mechanisms of action for electricity have been proposed for treatment of pain. *Firstly*, a segmental inhibition of pain signals to the brain in the dorsal horn of the spinal cord and *second*, activation of the descending inhibitory pathway with enhanced release of endogenous opioids and other neurochemical compounds (serotonin, noradrenaline, gamma aminobutyric acid (GABA), acetylcholine and adenosine).

Basically, electrotherapy is based on the following types: transcutaneous electrical nerve stimulation (TENS) and percutaneous electrical nerve stimulation (PENS). Instead of its pain-relieving effect, electrical stimulation has also been described for treatment of fecal and urinary incontinence, obesity, diabetes, postoperative ileus, chronic obstructive pulmonary disease, sarcopenia, symptomatic peripheral neuropathy in end-stage renal disease…

The aim of this paper is to achieve a better understanding of the different mechanism of action of both electrical stimulation techniques, especially for control of pain.

Figure 1. Artist's impression of use of electrical torpedo fish in the treatment of gout (a) and headache (b). Reproduced with permission from Perdikis 1977, South Africa Journal of Surgery [3].

Transcutaneous Electrical Nerve Stimulation (TENS) and Percutaneous Electrical Nerve Stimulation (PENS)

TENS is the acronym for *transcutaneous electrical nerve stimulation*. TENS is currently the most frequent non-invasive form of non-pharmacological pain management. This technique is based on the observation by William Sweet along with Wall [4] in 1967 when they stimulated their own infraorbital nerve by a needle electrode with 100 Hz, experiencing in a dramatic relief of pain. TENS consists of a battery-powered portable electric unit with electrodes applied to the skin, delivering electrical impulses to the underlying nerve fibers. TENS can be applied with either low frequency (LF-TENS < 10 Hz) or high frequency (HF-TENS 50-100 Hz) stimulation on the skin, but not at the same time. At high as well low frequencies TENS produces analgesia by activating smaller motor afferents while high frequency is more selective in stimulating larger diameter a beta afferents to cut down the nociceptor cell activity [5, 6].

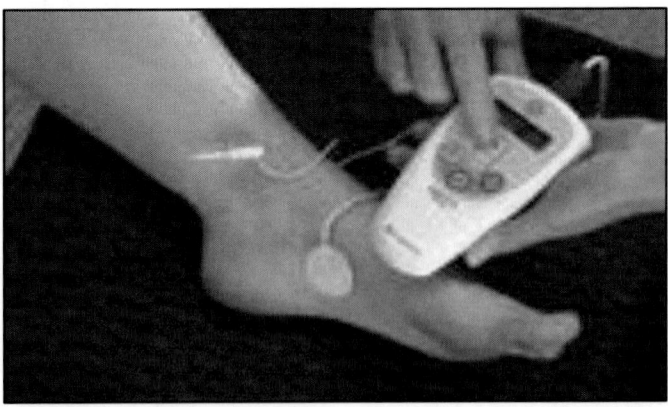

Figure 2. PENS at tibial posterior for treatment of fecal incontinence.

Users can self-administer TENS, adjusting the pulse amplitude (mA), frequency (pulses per second - pps), width or duration (μs) and pattern of the currents. Normally, there is no potential for overdose and there are few side effects or drug interactions. Effects are generally rapid in onset and offset so patients are encouraged to administer TENS as needed and throughout the day. TENS is cheap when compared to long term drug therapy [6].

On the other hand, *percutaneous electrical nerve stimulation* (PENS) is an hybrid technique combining TENS and acupuncture utilizing a needle electrode inserted in to soft tissues, according to Chinese landmarks of the body (trigger points), to modulate the small nerve endings in the soft tissues (Figure 2). PENS quickly alternates rhythm of high and low frequencies in order to achieve similar effects of stimulation as TENS [7]. PENS is usually indicated in cases of patients with intolerance to TENS (skin irritation, allodynia) and in conditions of skin resistance, in order to achieve the full potential of stimulation [8]. Moreover, it is also reported that PENS is more effective in post-herpetic neuropathy, severe sciatica and lower back pain (among others due to bone metastasis).

MECHANISM OF ACTION OF ELECTRICAL STIMULATION

The scientific bases of electrotherapy were based on the publication of Melzack and Wall's "Pain Mechanisms: A New Theory" [9] in 1965, in which they developed the Gate Control Theory of Pain. The authors suggested that "the substantia gelatinosa in the dorsal horn acts as a gate control system, which modulates the synaptic transmission of nerve impulses from peripheral fibers to the central cells." According to theory, the small nociceptive A-δ and C fibers maintain the hypothetical gate in a relative opened position, while stimulation of the large mechanoreceptive A-β fibers close the gate and inhibit the pain transmission to the brain. Because small nociceptive fibers are characterized by a higher threshold of action than the large mechanoreceptive fibers, the selective stimulation of the latter ones can prevent or reduce pain transmission (Figure 3). This hypothesis was the basis for the application of electrical stimulation to electrode bandaged to the skin, the aboved called TENS [10].

In addition to this mechanism of action, Melzack and Wall proposed that an activation of descending pain inhibitory systems is also implicated in the analgesic action of electricity. The explanation of this pathway was firstly exposed in experimental studies with rats by Woolf et al. [11]. They observed that the analgesic action of HF-TENS was partially reduced by cutting of the spinal cord of rats, due to the removing of the descending inhibitory components. Furthermore, Sluka et al. [12] reported that the analgesic effect of TENS outlasts the time of stimulation by several hours, implicating the involvement of extra-segmental factors.

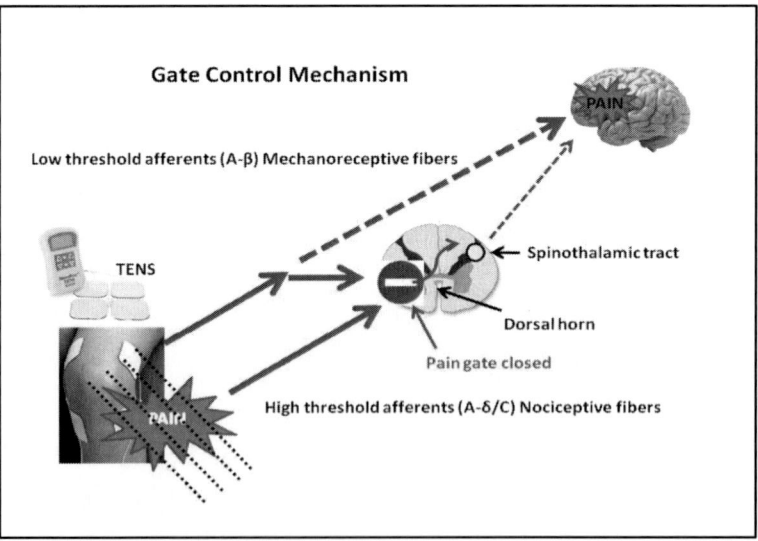

Figure 3. Gate of the pain mechanism. Activation of nociceptive fibers remains the gate open, while activation of mechanoreceptive fibers closes the gate and inhibits pain transmission to the brain (Modified from Heidland et al. Clinical Nephrology 2012 [2].

It is important to know that regarding frequency chosen (low or high), a different physiological or therapeutical effect is going to achieve. Anyway, current evidence suggests that in both frequencies, the secrection of endogenous opiods is the main physiological mechanism responsible of analgesia induced by TENS [13]. There are 3 different types of opiod receptors μ, λ, κ which are located peripherically, in spinal cord and supraspinal cord relationed with the descending pain inhibitory system. In each of these three levels, TENS presents different physiological mechanism of action (Figure 4).

Peripheral Physiological Mechanisms

Traditionally, it was known that analgesic effect of TENS was vehiculated only by activation of superficial cutaneous afferent nerve fibers. However, Tinazzi et al. [14] in 2005, observed that deep afferents also played an important role in analgesia induced by TENS. De Santana et al. [15] showed that only large mechanoreceptive A-β fibers located deeper were activated both with high and low frequency TENS.

Peripheral, at the beginning it was suggested that especially HF-TENS induced the analgesic effect because the nerve transmission of afferent fibers was fatigated or interrupted. However, a recently study by Ristic et al. [16] showed a decreasing of the range of evocate potential due to the antidromic collision of potentials produced by TENS in Aβ fibers.

Other study of King et al. [17] observed in rats that both HF-TENS and LF-TENS decreased their analgesic effect in those genetically manipulated animals in which the α2A adrenergic receptor was left. Furthermore, in those control rats, the administration of an antagonist of this α2A adrenergic receptor reverted the analgesic effect of TENS whether administration was peripherically inside articulation, without effect in case of intratecal or supraspinal administration.

Peripheral opioid receptors play an important role in analgesia produced by LF-TENS. Sabino et al. [18] showed in rats that µ-opioid receptros located peripherically participated in the analgesic effect of LF-TENS but not with the higher one.

On the other hand, it is suggested that adenosine can be implicated in the mechanism of action of TENS, because in the study of Marchand et al. [19] in humans, the administration of 200 mg of cafein (a blocker of adenosine receptors) previous to stimulation by TENS, significative reduced the analgesia in compare to placebo.

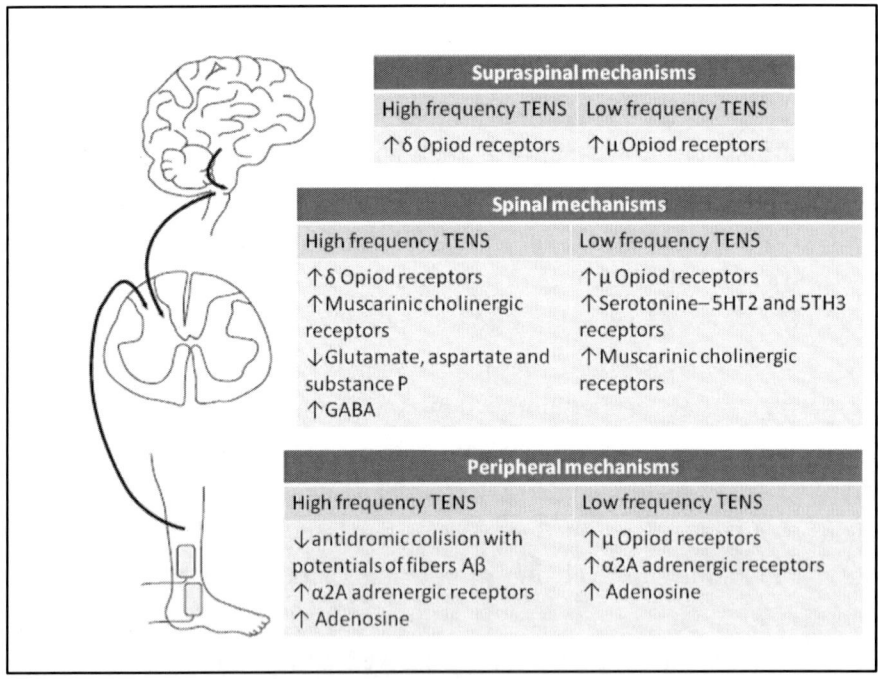

Figure 4. Physiological mechanisms related with pain activated by transcutaneous electrical nerve stimulation (TENS). Modified from J.J. Amer-Cuenca, C. Goicoechea and J.F. Lisón. ¿Qué respuesta fisiológica desencadena la aplicación de la técnica de estimulación nerviosa eléctrica transcutánea?. Rev Soc Esp Dolor. 2010;17(7):333–342 [13].

Spinal Physiological Mechanisms

Spinal effects of TENS are associated to the activation of at least 4 different kinds of receptors: opioids, serotoninergics, cholinergics and GABAergics [11, 20-24].

Regarding stimulation was with high or low frequency; different opioid receptors will be activated. In 1999, Sluka et al. [20] investigated the role of different types of opiod receptors μ, λ, κ in the analgesic mechanism of HF-TENS and LF-TENS. Low doses of naloxone, jointed to μ receptors, blocked the effect of LF-TENS, while blocking the λ opioid receptors prevent the effect of HF-TENS. On the other hand, blocking the κ opioid receptors did not produce any effect over analgesia neither higher nor lower frequency of TENS. This experiment was important because it could be showed that in spinal cord, LF-TENS acts mediated the μ-opioid receptors, meanwhile the λ-opioid receptors play the physiological effect of HF-TENS.

On the other hand, and after an articular inflammation, there is a physiologic rise of glutamate and aspartate. Moreover, it is already known that opioids reduce the secrection of glutamate and aspartate. Sluka et al. [25] in 2005 conducted another study in which observed that HF-TENS, but not LF-TENS, significative reduced liberation of glutamate and aspartate in animals with articular inflammation in compare with animals without inflammation. This descense was not produced wether a previous administration of naltrindiol (λ-opiod receptor antagonist) was administrated.

In 2006, Sluka et al. [26] performed a study with arthritis rats in order to determine wether serotonin and noreadrenaline influenciate in the mechanism of action of TENS. They observed an increase of serotonin levels in the dorsal hord of spinal cord when LF-TENS were applicated, but no changes were observed when HF-TENS were applicated. These results confirm those obtained by Radhakrishnan et al. [23] in which the block of serotonin receptors 5-HT2 y 5-HT3 avoided the analgesic effect of the LF-TENS, but not the high frequency ones. In case of noradrenaline levels, no changes were found in the dorsal hord of spinal cord neither application of lower nor higher frequency TENS, so in that sense, it seems that noradrenergics receptors do not act in the mechanism of action of TENS [21].

Cholinergic's receptors are also implicated in the mechanism of action of TENS. These receptors are located in the dorsal hord of spinal cord and can be muscarinic and nicotinic. In 2003, Radhakrishnan et al. [21], observed the effect of both low and high frequency TENS were related to muscarinic receptors instead of nicotinic ones.

Recently, it has also implicated the paper of gamma aminobutyric acid (GABA). GABA is the principal neuroinhibitory transmitter, and it is segregated in the deep dorsal horn of the spinal cord. Maeda et al. [24], showed that administration of HF-TENS increases concentrations of GABA in the spinal cord of rats with and without inflammation. However, this enhance of GABA was not produced whether application of frequency was low. In contrast, if an antagonist of GABA receptors (bicuculine) was administrated, a reduction of the effect of both low and high frequency was observed.

Finally, TENS has also relationed with changes in the P substance in this region. P substance is implicated in painful mechanisms and it has been observed that administration of antagonists to this substance in the spinal cord, produces an analgesic effect [27].

Supraspinal Physiological Mechanisms

In the '80s, several studies [28-30] observed that stimulation with TENS produce an increase of β-endorphines at blood plasma and spinal fluid in humans. Due to the studies of DeSantana et al. [15, 31], it is already known that the descending pathway originates in the midbrain and starts in the periaqueductal grey (PAG), which sends projections to the rostral ventral medulla (RVM), followed by projections to the spinal dorsal horn. They observed that TENS-induced activation of PAG and RVM is involved in the reduction of hyperalgesia in arthritic rats [15]. It is also important the findings of Sabino et al. [18] about the enhanced release of endogenous opioids and serotonin, which act through the PAG-RVM pathway. A classical study of Kalra et al. [22] analyzed the spinal fluid of arthritic rats and observed an enhanced of the levels of μ-opiods in those rates treated with low frequency of TENS; and an increase of the concentrations of δ-opiods in the high frequency group. When a preoperative treatment of these rats with the μ-opioid receptor antagonist, naloxone, a blocked of the effect of low frequency TENS was observed; while the administration of the δ-opioid receptor antagonist, naltrindole, prevented the action of high frequency TENS.

Non-Analgesics Peripheric Physiologic Mechanisms

According to laser Doppler investigations, TENS application stimulates the peripheral microcirculation, potentially contributing to the analgesic action of TENS [32]. In diabetic neuropathy, a relationship between capillary abnormalities and severity of neuropathy has been observed [33]. In these patients TENS-induced vasodilation is associated with the enhanced microcirculation and increased endoneural blood flow [34].

Another effect could be the alteration in nerve transmission. However, the results are not consistent and studies have not found significative changes in nerve afferents [14, 35, 36].

Finally, TENS can also use for neuromuscular stimulation in order to achieve muscular contractions (neuromuscular electrical stimulation: NMES). Some studies have measured the electromiographic activity and the force with dynamometers before and after application of TENS, and they have observed an improve effect over muscular activity. Conclusion of these studies could be the role of TENS in patients with motor deficits such as cerebrovascular events or other neurological disorders [14, 37, 38].

REFERENCES

[1] Rossi U. The history of electrical stimulation of the nervous system for the control of pain. *Electrical stimulation and the relief of pain.* Edited by Simpson BA, Elsevier BV 2003. p. 6 – 16.
[2] Heidland A, Fazeli G, Klassen A, Sebekova K, Hennemann H, Bahner U and Di Iorio B. Neuromuscular electrostimulation techniques: historical aspects and current possibilities in treatment of pain and muscle wasting. *Clinical Nephrology*, Vol. 78 – No. Suppl. 1/2012 (S12-S23).
[3] Perdikis P. Transcutaneous nerve stimulation in the treatment of protracted ileus. *S Afr J Surg.*1977; 15: 81-86.
[4] Wall PD, Sweet WH. Temporary abolition of pain in man. *Science.* 1967; 155: 108-109.
[5] Perryman LT. Peripheral Nerve Stimulation and Percutaneous Electrical Nerve Stimulation in Pain Management: A Review and Update on Current Status. *Int J Pain Relief.* 2017;1(1): 036-041.
[6] Johnson MI. Transcutaneous electrical nerve stimulation. In: Kitchen S, ed. *Electrotherapy: Evidence Based Practice.* Edinburgh: Churchill Livingstone; 2002: 259-286.
[7] Ghoname ES, Craig WF, White PF, Ahmed HE, Hamza MA, Gajraj NM, et al. The effect of stimulus frequency on the analgesic response to percutaneous electrical nerve stimulation in patients with chronic low back pain. *Anesth Analg.* 1999; 88: 841-846.
[8] Raphael JH, Raheem TA, Southall JL, Bennett A, Ashford RL, Williams S. Randomized double-blind sham-controlled crossover

study of short-term effect of percutaneous electrical nerve stimulation in neuropathic pain. *Pain Med.* 2011; 12: 1515-1522.
[9] Melzack RA, Wall PD. Pain mechanisms: a new theory. *Science.* 1965; 150: 971-979.
[10] Gildenberg PL. History of Electrical neuromodulation of chronic pain. *Pain Medicine.* 2006; 7: S7-S13.
[11] Woolf CJ, Mitchell D, Barrett GD. Antinociceptive effect of peripheral segmental electrical stimulation in the rat. *Pain.* 1980; 8: 237-252.
[12] Sluka KA, Walsh D. Transcutaneous electrical nerve stimulation: basic science mechanisms and clinical effectiveness. *J Pain.* 2003; 4: 109-121.
[13] Amer-Cuenca JJ, Goicoechea C and Lisón JF. ¿Qué respuesta fisiológica desencadena la aplicación de la técnica de estimulación nerviosa eléctrica transcutánea?. *Rev Soc Esp Dolor.* 2010; 17(7):333–342.
[14] Tinazzi M, Zarattini S, Valeriani M, Romito S, Farina S, Moretto G, et al. Long-lasting modulation of human motor cortex following prolonged transcutaneous electrical nerve stimulation (TENS) of forearm muscles: evidence of reciprocal inhibition and facilitation. *Exp Brain Res.* 2005; 161:457–464.
[15] DeSantana JM, Walsh DM, Vance C, Rakel BA, Sluka KA. Effectiveness of transcutaneous electrical nerve stimulation for treatment of hyperalgesia and pain. *Curr Rheumatol Rep.* 2008; 10: 492-499.
[16] Ristic D, Spangenberg P, Ellrich J. Analgesic and antinociceptive effects of peripheral nerve neurostimulation in an advanced human experimental model. *Eur J Pain.* 2008;12:480–490.
[17] King EW, Audette K, Athman GA, Nguyen HO, Sluka KA, Fairbanks CA. Transcutaneous electrical nerve stimulation activates peripherally located alpha-2A adrenergic receptors. *Pain.* 2005; 115:364–373.

[18] Sabino GS, Santos CM, Francischi JN, de Resende MA. Release of endogenous opioids following transcutaneous electric nerve stimulation in an experimental model of acute inflammatory pain. *J Pain*. 2008; 9: 157-163.
[19] Marchand S, Li J, Charest J. Effects of caffeine on analgesia from transcutaneous electrical nerve stimulation. *N Engl J Med*. 1995; 333:325–326.
[20] Sluka KA, Deacon M, Stibal A, Strissel S, Terpstra A. Spinal blockade of opioid receptors prevents the analgesia produced by TENS in arthritic rats. *J Pharmacol Exp Ther*. 1999; 289: 840–846.
[21] Radhakrishnan R, Sluka KA. Spinal muscarinic receptors are activated during low or high frequency TENS-induced antihyperalgesia in rats. *Neuropharmacology*. 2003;45:1111–1119.
[22] Kalra A, Urban MO, Sluka KA. Blockade of opioid receptors in rostral ventral medulla prevents antihyperalgesia produced by transcutaneous electrical nerve stimulation (TENS). *J Pharmacol Exp Ther*. 2001; 298: 257-263.
[23] Radhakrishnan R, King EW, Dickman JK, Herold CA, Johnston NF, Spurgin ML, et al. Spinal 5-HT(2) and 5-HT(3) receptors mediate low, but not high, frequency TENS-induced antihyperalgesia in rats. *Pain*. 2003;105:205–213.
[24] Maeda Y, Lisi TL, Vance CGT, Sluka KA. Release of GABA and activation of GABA (A) in the spinal cord mediates the effects of TENS in rats. *Brain Res*. 2007;1136:43–50.
[25] Sluka KA, Vance CGT, Lisi TL. High-frequency, but not lowfrequency, transcutaneous electrical nerve stimulation reduces aspartate and glutamate release in the spinal cord dorsal horn. *J Neurochem*. 2005; 95:1794–1801.
[26] Sluka KA, Lisi TL, Westlund KN. Increased release of serotonin in the spinal cord during low, but not high, frequency transcutaneous electric nerve stimulation in rats with joint inflammation. *Arch Phys Med Rehabil*. 2006; 87:1137–1140.

[27] Doubell TP, Mannion RJ, Woolf CJ. The dorsal horn: Statedependent sensory processing, plasticity and the generation of pain. En: Wall PD, Melzack R, editores. *Textbook of Pain*, 4th ed. Philadelphia: Churcill Livingstone; 2003. p. 165–182.

[28] Salar G, Job I, Mingrino S, Bosio A, Trabucchi M. Effect of transcutaneous electrotherapy on CSF beta-endorphin content in patients without pain problems. *Pain*. 1981; 10: 169-172.

[29] Hughes GS Jr, Lichstein PR, Whitlock D, Harker C. Response of plasma beta-endorphins to transcutaneous electrical nerve stimulation in healthy subjects. *Phys Ther*. 1984; 64: 1062-1066.

[30] Fields HL, Basbaum AL. Central nervous system mechanisms of pain modulation. En: Wall PD, Melzack R, editores. *Textbook of Pain*, 4th ed. Philadelphia: Churchill Livingstone; 2003. p. 243–257.

[31] DeSantana JM, Da Silva LF, De Resende MA, Sluka KA. Transcutaneous electrical nerve stimulation at both high and low frequencies activates ventrolateral periaqueductal grey to decrease mechanical hyperalgesia in arthritic rats. *Neuroscience*. 2009; 163: 1233-1241.

[32] Wikström SO, Svedman P, Svensson H, Tanweer AS; Sven Olof Wikström, Paul Svedman, H. Effect of transcutaneous nerve stimulation on microcirculation in intact skin and blister wounds in healthy volunteers. *Scand J Plast Reconstr Surg Hand Surg*. 1999; 33: 195-201.

[33] Malik RA, Newrick PG, Sharma AK, Jennings A, Ah-See AK, Mayhew TM, Jakubowski J, Boulton AJ, Ward JD. Microangiopathy in human diabetic neuropathy: relationship between capillary abnormalities and the severity of neuropathy. *Diabetologia*.1989; 32: 92-102.

[34] Kaada B. Vasodilation induced by transcutaneous nerve stimulation in peripheral ischemia (Raynaud's phenomenon and diabetic polyneuropathy). *Eur Heart J*. 1982; 3: 303-314.

[35] Fernández Del Olmo M, Álvarez-Sauco M, Koch G, Franca M, Márquez G, Sánchez JA, et al. How repeatable are the physiological effects of TENS? *Clin Neurophysiol.* 2008; 119: 1834–1839.

[36] Alves-Guerreiro J, Noble JG, Lowe AS, Walsh DM. The effect of three electrotherapeutic modalities upon peripheral nerve conduction and mechanical pain threshold. *Clin Physiol.* 2001; 21:704–711.

[37] Koesler IB, Dafotakis M, Ameli M, Fink GR, Nowak DA. Electrical somatosensory stimulation improves movement kinematics of the affected hand following stroke. *J Neurol Neurosurg Psychiatry.* 2009; 80:614–619.

[38] Yan T, Hui-Chan CW. Transcutaneous electrical stimulation on acupuncture points improves muscle function in subjects after acute stroke: a randomized controlled trial. *J Rehabil Med.* 2009;41:312–316.

In: Uses of Electrical Stimulation... ISBN: 978-1-53615-036-0
Editor: Jaime Ruiz-Tovar © 2019 Nova Science Publishers, Inc.

Chapter 2

A PERCUTANEOUS ELECTRICAL NEUROSTIMULATION OF THE POSTERIOR TIBIAL NERVE FOR FECAL INCONTINENCE

Maria del Mar Aguilar-Martínez, MD,
Luis Sánchez-Guillén*, MD
and Antonio Arroyo, MD, PhD

Colorectal Surgery Unit. Service of General and Digestive Surgery.
University General Hospital of Elche, Alicante, Spain
Department of Pathology and Surgery, Miguel Hernández University
of Elche, Elche, Alicante, Spain

ABSTRACT

Fecal incontinence (FI) is a prevalent pathology with a great impact on the quality of life of patients. Despite the treatment options, there are no standardized regimens available. The neurostimulation of the posterior tibial nerve (PTNS) has been imposed on other treatments as a less invasive and more economical alternative, with a clinical and manometric

---

[*] Corresponding Author Email: drsanchezguillen@gmail.com.

improvement in the short-medium term. In addition, recent studies indicate that the continuity of treatment with sessions more spaced over time, offers long-term benefits.

Keywords: tibial, fecal, incontinence, percutaneous

INTRODUCTION

(FI) is defined as the partial or total loss of ability to voluntarily control the expulsion of gases and fecal material. It can be associated with urinary incontinence (UI) in 8 - 28% [1].

The estimated prevalence of FI is 7.7%, increasing with age up to 10% in people over 65 years. Regarding sex, the incidence is similar in men (8.1%) and in women (8.9%) [2]. In patients admitted to geriatric institutions FI is more frequent, reaching up to 50%.

The pathogenesis of FI is wide, and there are predisposing and causal factors such as chronic constipation, inflammatory bowel disease, neuropathies, degenerative disorders, iatrogenesis and even idiopathic pathology.

The FI supposes a deterioration in the quality of life of patients, especially in severe cases, associating an important physical and psychological comorbidity. However, despite the negative impact on quality of life and the possibility of treatment, only 10 - 30% of patients consult the specialist, which is why it is an underdiagnosed pathology [3].

There are several treatment options available. Conservative management with hygienic-dietary habits and biofeedback is effective in 50% of cases, especially mild ones. Surgery is useful in cases of FI due to anatomical anomalies or serious and recent injuries of the sphincters, with different techniques depending on the cause of FI (sphincteroplasty, total pelvic floor repair, muscle transposition). However, all interventions are invasive and generally, in 50% of cases, continence disorders can be observed.

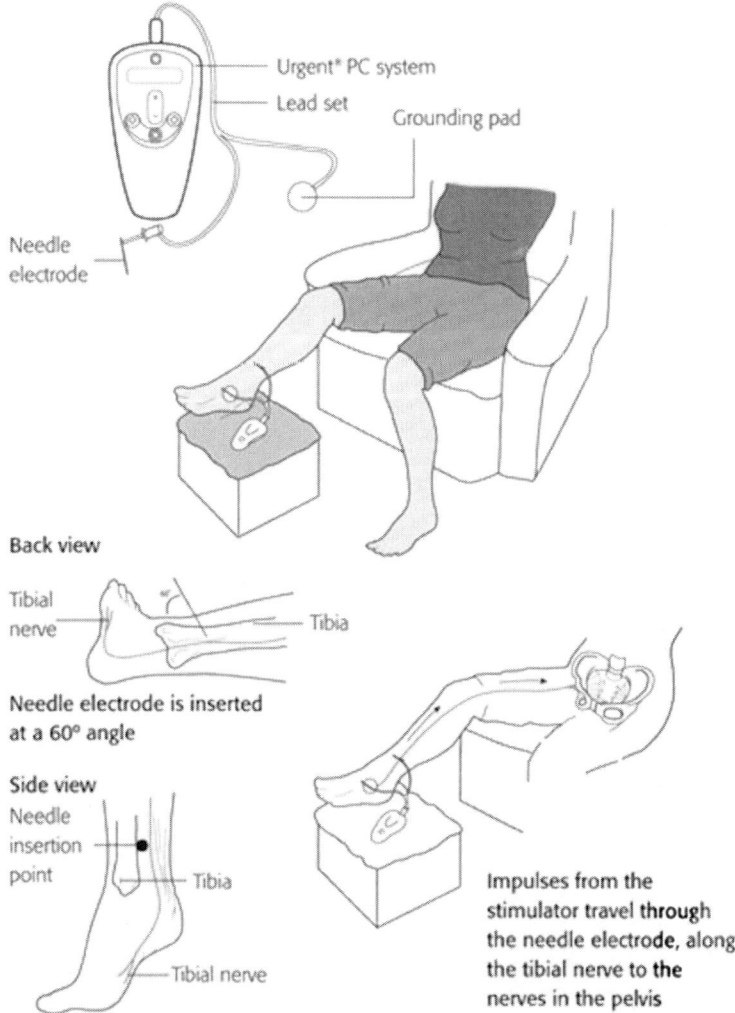

Figure 1. Percutaneous tibial nerve stimulation [10].

Neurostimulation appears as a new alternative for the treatment of patients with fecal incontinence without surgical options and in whom biofeedback has failed [4].

The first case of fecal incontinence treated with neurostimulation, was 20 years ago and it was in the variant of sacral stimulation. It has been shown that the external sphincter is innervated by nerve from the sacral roots S3 and S4. The stimulation of them produces continuous contractions in the sphincter, increasing the tone and contraction capacity [5]. However, it is a high-cost, invasive technique with some complications, such as mobilization of the electrode, which requires a new intervention, or local pain related to the generator.

In the search for a less invasive, safe, effective and lower cost technique, posterior tibial neurostimulation (PTNS) arises.

PTNS is based on electrical stimulation of the posterior tibial nerve. This nerve contains fibers originated in the L4-L5 roots as well as in the sacral S1, S2 and S3, so that its activation reduces the rates of fecal incontinence. PTNS is performed percutaneously the upper the internal malleolus by means of an electrode and needle, being a safe and economical technique [6].

Mechanism of Action

The first neurostimulation publications are from 1995 and are related to results obtained from the neurostimulation of sacral nerves, with success rates up to 90% [7, 8]. However, the rate of adverse effects reaches 12.8% [8].

PTNS was initially described as a treatment for urological disorders and was later adapted to the treatment of fecal incontinence by the group of Shafik et al. [9]. This technique produces, in a minimally invasive way, a neuromodulation of the spinal roots L4-S3, responsible for the innervation of the pelvic floor [10]. Although the mechanism of action is not completely known, it is thought that stimulation of the tibial nerve achieves that, in the form of a somato-autonomic reflex arc, the roots of the sacral plexus are stimulated in relation to the visceral and muscular control of the pelvic floor.

Its effect would include a stimulation of the striated musculature, with a consequent increase in the maximum pressure of contraction and rest. There is also evidence of a reduction in both, spontaneous anal relaxation and rectal contraction. On the other hand, there would be an increase in the blood flow of the rectal mucosa (which controls autonomic nervous function) [11].

METHODS

Participants

The patient for treatment with PTNS, is one in which have failed, both conservative measures and biofeedback.

In the outpatient clinic, and before starting therapy, they were made a physical examination, analyze the Wexner scale value and perform an ultrasound and anorectal manometry.

PTNS

Neurostimulation should be performed by colorectal surgeons, although it could also be carried out by qualified nurses.

There are two available therapies to carry out neurostimulation: transcutaneous and percutaneous.

Percutaneous Therapy

The punction place is inserted approximately 5 cm cephalad to the medial malleolus and posterior to the tibia. The posterior tibial nerve is stimulated by a conductor that is inserted between 1-3 mm deep into the skin and with a surface electrode on the arch of the foot. Energy is applied progressively with a neurostimulator from the company Uroplasty, model Urgent PC stimulator ® until stimulation is achieved.

The patient may have flexion of the distal phalanx of the first finger or a stimulus that radiates to the first metatarsal and to the groin.

In the Urgent PC ® neurostimulator, there are 20 intensity levels that generate an electrical amplitude between 0.5 - 9 mA. According to the requirements of the patient, the intensity is increased or decreased.

The percutaneous stimulation of the posterior tibial nerve, is a novel therapy in the treatment of FI, which is still under study.

	Never	Rarely	Sometimes	Usually	Always
Solid	0	1	2	3	4
Liquid	0	1	2	3	4
Gas	0	1	2	3	4
Wears pad	0	1	2	3	4
Lifestyle alteration	0	1	2	3	4

Figure 2. Wexner Score.

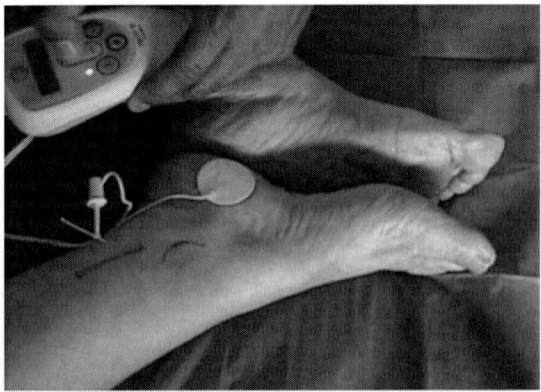

Figure 3. External neurostimulator and point of application of PTNS.

Actually, in the scientific literature, only 12 studies, one clinical trial and one review have been published. In addition, the follow-up periods of these jobs are short, taking an average of 6 months.

There is no doubt about the efficacy of PTNS in the short term, which is comparable to neurostimulation of sacral nerves. However, it is still pending to analyze its efficacy in the medium and long term [4].

Transcutaneous Therapy

The posterior tibial nerve neurostimulation in this case is performed with the use of a TENS (Transcutaneous electrical nerve stimulation).

The stimulation is performed with surface electrodes, there are two types: the negative electrode, which is placed at the dorsal ankle to the internal malleolus, and the positive electrode, which is placed 10 cm above the negative electrode. The proper position of the electrode is determined by the visualization of the flexor response of the toes during the stimulation. The ideal intensity level corresponds to the lower limit of the threshold of motor contraction and ranges between 10 and 35 mA.

The stimulation is applied for 20 minutes, 5 days a week for a month.

The transcutaneous therapy is less investigated than the percutaneous one. There are few studies that analyze it and only show advantages from the point of view of the symptomatology derived from the punction place [4].

Scheme of Treatment

Patients attend sessions of 30 minutes, with a total of 12 sessions in three months in the first phase, 6 sessions in 3 months in the second phase and 6 sessions in 6 months in the third and final phase.

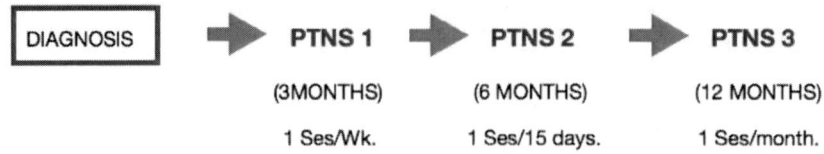

	DAY						
Using tally marks, mark the following boxes each time one of the following situations occurs:							
	1	2	3	4	5	6	7
1. I ran to the bathroom, but I MADE IT IN TIME and did not have an accident.							
2. I ran to the bathroom, but I HAD A FAECAL ACCIDENT (partial or total).							
3. I had a faecal accident without realising it.							
4. I had a normal bowel movement WITHOUT HAVING TO RUN.							
TOTAL NUMBER OF BOWEL MOVEMENTS (add the tally marks from numbers 1, 2, 3 and 4)							

	DAY						
At the end of the day, it is important to answer each question (circle the answer)							
	1	2	3	4	5	6	7
Are you wearing an incontinence pad today?	YES NO	YES NO	YES NO	YES NO	YES NO	YES NO	YES NO
Did you soil the pad or your underwear today?	YES NO	YES NO	YES NO	YES NO	YES NO	YES NO	YES NO
What was your stool like today?	Hard Normal Soft	Hard Normal Soft	Hard Normal Soft	Hard Normal Soft	Hard Normal Soft	Hard Normal Soft	Hard Normal Soft
Did you take a laxative today? Write which one.	YES NO	YES NO	YES NO	YES NO	YES NO	YES NO	YES NO
Did you take something to bind your stools today? Write what you have taken.	YES NO	YES NO	YES NO	YES NO	YES NO	YES NO	YES NO
Have you done any manoeuvres to improve continence today (enemas, suppositories, etc.)?	YES NO	YES NO	YES NO	YES NO	YES NO	YES NO	YES NO
Has faecal leakage affected your social, sexual or work-related activities today?	YES NO	YES NO	YES NO	YES NO	YES NO	YES NO	YES NO

Figure 4. Defecation diary [12].

During the treatment, to assess the severity and the number of average bowel movements per week, the continence diary is used, which establishes the severity of the IF in three stages: mild (from 3 to 6 episodes a week), moderate (from 7 to 11) and severe (more than 11). (Image 4)

According to the number of leaks I feel

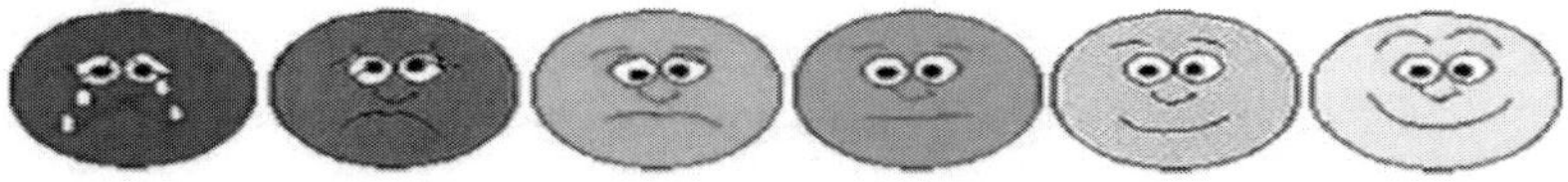

You are very bad	You are bad	You are regular	You are well	You are very well	You are excellent
10	8	6	4	2	0

Figure 5. Rapid assessment faecal incontinence score [13].

Quality of life is assessed using the Life Quality Scale adapted to the FI (FIQLs) whose values for the 4 items range from 1 to 4. The higher the score is the better health status. (Image 5).

At the end of the last phase of treatment, the patient goes to the clinic.

During this visit, a physical examination was performed, assessment of the improvement of incontinence, comparing the value obtained in the Wexner scale at the beginning of the treatment and also, it is performed an ultrasound and anorectal manometry.

Results

In the literature, treatment success is defined as the 50% reduction of incontinence episodes per week.

Thus, the success rate of PTNS in patients with FI is very high, being 72.5%. The most benefited patients are those who present urgency incontinence.

Actually, data analysis of the life quality questionnaires and the incontinence diary continues in controversy due to the subjectivity of the same. However, longer series indicate that between 60 - 70% experience improvement reflected in the incontinence diary and quality of life questionnaires [14, 15].

Regarding manometry, the results are not conclusive. The maximum basal pressure that usually starts with values around 16 mmHg, becomes at the end of the treatment of 24.2 mm Hg ($p > 0.01$). However, the maximum squeeze pressure with the effort does obtain more solid results, reaching up to 209 mmHg [14].

In addition to the manometry, it is included in some works other more objective data, the evaluation of the capacity of contain. At the beginning of treatment with PTNS, 62.8% of patients are not able to contain feces more than one minute. Once the three phases of stimulation are finished, 56.2% are able to retain them more than ten minutes [15].

Complications

The complications derived from PTNS are infrequent, having an overall incidence of 1 - 2%.

Mild complications can be described, including bleeding at the punction place, hematoma or pain; but also more complex such as paresthesias or abdominal pain.

In the multicentre study conducted by Govaert, the complication rate is 16.63%. Paresthesias are the most frequent complication. Normally, they are self-limited, with less than 2 hours and that usually do not require the suspension of treatment. Abdominal pain is another complication that can appear. In the same way as it occurs with paresthesias, it is self-limited, approximately 2 hours in duration and does not require the suspension of treatment [16].

Future and Dilemmas of PTNS

Although PTNS is effective in many studies for the treatment of FI, there are still some dilemmas.

The treatment protocol most accepted by the scientific community is that of 30-minute sessions, for 2 weeks. However, the results that exist with this protocol are short-medium term, which creates uncertainty regarding the efficacy and the correct duration of the treatment. There are some studies in which it is demonstrated that the continuity of the treatment with sessions more spaced over time offers long-term benefits. However, more studies are needed to analyze whether it is necessary to increase the number of sessions, in order to provide a stable Wexner throughout life [4].

Regarding unilateral or bilateral stimulation, there are no studies comparing the results of both pathways. A small recent work demonstrates greater benefit with bilateral transcutaneous stimulation, compared with unilateral percutaneous. However, the sample is small and larger studies are needed to confirm these results [4].

REFERENCES

[1] Wenzel, C., Rondimi, C., Troncoso, R., et al. (2009). Prevalence os fecal incontinence in gynecological and urogenecological patients. *Rev Chil Obstet Ginecol*, 74 (VI): 354–359.

[2] Ng, KS., Sivakumaran, Y., Nassar, N., et al. (2015). Fecal Incontinence: Community Prevalence and Associated Factors. A Systematic Review. *Dis Colon Rectum*, 58 (XII): 1194-209.

[3] Dunivan, G., Heymen, S., Palsson, O., et al. (2010). Fecal incontinence in primary care: prevalence, diagnosis, and health care utilization. *Am J Obstet Gynecol*, 202 (V): 493-e1.

[4] George, A., Maitra, R., Maxwell-Armstrong, C., (2013). Posterior tibial nerve stimulation for fecal incontinence: Where are we?. *World J Gastroenterol*, 19 (XLVIII): 9139–9145.

[5] Dudding, TC., Parés, D., Vaizey, CJ. (2009). Comparison of clinical outcome between open and percutaneous lead insertion for

permanent sacral nerve neurostimulation for the treatment of fecal incontinence. *Dis Colon Rectum*, 52 (III): 463-468.

[6] Hotouras, A., Murphy, J., Allison, M., et al. (2014). Prospective clinical audit of two neuromodulatory treatments for fecal incontinence: sacral nerve stimulation (SNS) and percutaneous tibial nerve stimulation (PTNS). *Surg Today*, 44 (XI): 2124-2130.

[7] Simillis, C., Lal, N., Qiu, S., et al. (2018). Sacral nerve stimulation versus percutaneous tibial nerve stimulation for faecal incontinence: a systematic review and meta-analysis. *Int J Colorectal Dis*, 33 (V): 645-648.

[8] Matzel, K., Stadelmaie, U., Gall, F., et al. (1995). Electrical stimulation of sacral spinal nerves for treatment of faecal incontinence. *Lancet*, 346 (8983): 1124-1127.

[9] Shafik, A., Ahmed, I., El-Sibai, O., et al. (2003). Percutaneous peripheral neuromodulation in the treatment of fecal incontinence. *Eur Surg Res*, 35 (II): 103–107.

[10] *Patient Pictures.* Clinical Drawing for patients. Percutaneous tibial nervestimulation. http://www.patientpictures.com/view.php?name =Percutaneous%20tibial%20nerve%20stimulation&cat=Urology %20and%20genitourinary%20medicine&id=326&src.

[11] Nakamura, M., Sakurai, T., Tsujimoto, Y., et al. (1983). Transcutaneous electrical stimulation for the control of frequency and urge incontinence. *Hinyokika Kiyo,* 29 (IX): 1053- 1059.

[12] Muñoz, A., Lagares, L., Vargas, H., et al. (2017). High-resolution circuit for the diagnosis of faecal incontinence. Patient satisfaction. *Cir Esp*, 95 (V): 276-282.

[13] De la Portilla, F., Calero, A., Jiménez, R., et al. (2015). Validation of a new scoring system: Rapid assessment faecal incontinence score. *World J Gastrointest Surg*, 7 (IX): 203-207.

[14] De La Portilla, F., Rada, R., Vega, J., et al. (2009). Evaluation of the use of posterior tibial nerve stimulation for the treatment of fecal incontinence: Preliminary results of a prospective study. *Dis Colon Rectum*, 52 (VIII): 1427–1433.

[15] Delgado, A., Arroyo, A., Ruiz-Tovar, J., et al. (2014). Effect on anal pressure of percutaneous posterior tibial nerve stimulation for faecal incontinence. *Colorectal Dis*, 16 (VII): 533-537.

[16] Govaert, B., Pares, D., Delgado, S., et al. (2010). A prospective multicentre study to investigate percutaneous tibial nerve stimulation for the treatment of faecal incontinence. *Colorectal Dis*, 12 (XII), 1236-1241.

In: Uses of Electrical Stimulation…   ISBN: 978-1-53615-036-0
Editor: Jaime Ruiz-Tovar   © 2019 Nova Science Publishers, Inc.

Chapter 3

PERCUTANEOUS ELECTRICAL NEUROSTIMULATION OF DERMATOME T6 FOR THE TREATMENT OF OBESITY

Jaime Ruiz-Tovar[*] *and Carolina Llavero*
Electrostimulation Unit. Clinica Garcilaso, Madrid, Spain

ABSTRACT

PENS of dermatome T6 is actually a consolidated therapy for weight loss. The participants undergo one 30-min session every week for 12 consecutive weeks. It has been demonstrated to be useful for mild to moderate obese patients in order to reduce a significant amount of weight and achieving many of them their normal weight. The effect of this therapy does not just end after finishing the last stimulation, but its anorexygenic effect continues at least during 3 months more. PENS of dermatome T6 must be associated to a hypocaloric diet. Thus, the survey by a dietician is essential, helping the patients to acquire healthy alimentary habits, and avoiding a weight regain once the effect get lost. One-year follow up studies have demonstrated that there is no weight regain, once the therapy has finished. The effect of appetite reduction is associated with a decrease in plasma levels of ghrelin.

---

[*] Corresponding author: Jaime Ruiz-Tovar, MD, PhD, E-mail: jruiztovar@gmail.com.

Keywords: percutaneous electrical neurostimulation, dermatome T6, weight loss, appetite reduction, dietary compliance, long-term effect, ghrelin

INTRODUCTION

About one third of the population in developed countries is obese to some degree. Obesity itself is a health risk factor that influences the development and progression of various diseases, such as dyslipidemia, ischemic heart disease, hypertension, type 2 diabetes mellitus and sleep apnea-hypopnea syndrome, thereby worsening the quality of life of patients, limiting their activities and causing psychosocial problems. There is a direct relationship between body mass index (BMI) and morbidity and mortality risks in obese patients, which is derived from associated pathologies and results in making obesity itself a disease [1-3].

Dietary treatment associated with physical exercise is the first therapeutic step for obesity. However, to be effective, patient motivation is essential, though often lacking. Obese patients often tire of following a low-calorie diet for long time periods. A continuous feeling of hunger is the major cause of dietary treatment failure. Another problem associated with diets is that once abandoned them, the patient often regains the previously lost weight [1, 4].

Finally, the last step in the treatment options for obesity is bariatric surgery. Surgery is the best method to obtain a significant and maintained weight loss. However, bariatric surgery has clear indications (patients with BMI > 40 Kg/m^2 or BMI > 35 Kg/m^2 associated to obesity-related comorbidities). Though actually most techniques are performed by laparoscopic approach, obtaining very low morbidity and mortality rates, this therapy remains the most risky option and therefore, it is reserved for failures of dietary treatment [5-7].

THEORETICAL BASIS FOR THE DEVELOPMENT OF PERCUTANEOUS ELECTRICAL NEUROSTIMULATION OF DERMATOME T6 TO REDUCE APPETITE

The initial development of this technique was based on 4 keypoints: the effect of the gastric pacemaker, the percutaneous electrical neurostimulation (PENS) of the posterior tibial nerve, the observations of weight loss after central or peripherical neural stimulation and the anatomical background:

Gastric Pacemaker

An implantable gastric stimulator (gastric pacemaker) has been used to treat obesity, with promising results. It applies cyclic electric pulses of 40 Hz every 4-12 minutes to the gastric wall. The stimulator induces gastric distention in the fasting state and inhibits postprandial antral contractions, thereby impairing stomach emptying, which may lead to early satiety and reduced food intake. The induction of gastric distension in the fasting state results in the activation of stretch receptors, causing satiety [7]. It has been observed that this technique achieves excess weight loss up to 40% after 1 year. This stimulator can be placed laparoscopically or endoscopically; both techniques imply a small risk to the patient, but are still invasive [8, 9]. The modulation of neuronal activities and release of certain hormones with an implantable gastric stimulator may also explain the reduction of appetite and the increase of satiety. A decrease in Ghrelin levels could be one mechanism that explains weight loss and appetite reduction after implantable gastric stimulation [10]. Chen reported that electric gastric stimulation with a gastric pacemaker may affect the central nervous system by segregating hormones in the stomach and regulating satiety and/or appetite, with ghrelin being particularly involved in this process [11].

Percutaneous Electrical Neurostimulation of the Posterior Tibial Nerve

Percutaneous electroneurostimulation (PENS) was originally developed to treat urinary and fecal incontinence by stimulating the posterior tibial nerve. The mechanism of action involves the creation of a somato-somatic reflex: the posterior tibial nerve is the afferent pathway, conducting the electrical impulse to root S3, and the efferent pathway is the pudendal nerve, which is responsible for the innervation of the anal sphincter [11, 12].

Weight Loss after Central or Peripherical Neural Stimulation

Pereira and Foster [13] observed an excess weight loss of 20% associated with decreased appetite in 2 morbidly obese patients, in whom spinal cord stimulators were set up at the T6 and T7 levels, to control intractable lumbar pain and lumbosacral radiculitis secondary to lumbar disc herniation. These patients did not increase their physical activity or follow any type of diet; however, they experienced significant appetite reductions. These authors were the first to hypothesize that spinal cord stimulation could affect the stomach. Other authors have reported that transcutaneous electrical gastric stimulation may alter gastric motility, delay gastric emptying and lead to postprandial satiety [14-16]. They believed that electrical stimulation was transmitted to the stomach through the abdominal wall when placing the electrode in the left upper quadrant of the abdomen. However, we think that it is more likely that the effect is produced by the creation of a somato-autonomic reflex rather than by transcutaneous transmission of the electrical stimuli, similarly to the transcutaneous electrical stimulation of the posterior tibial nerve in the incontinence treatment [17]. Moreover, it is difficult to believe that electrical stimulus could have some effect when traversing a thick abdominal wall, which is present in morbidly obese patients, particularly considering the presence of adipose tissue, which is not a good electrical conductor. The same authors have also postulated that the effect of gastric stimulation, associated with the delay of gastric emptying, might also decrease ghrelin segregation in the gastric fundus and inhibit appetite through the central nervous system [14-16].

Anatomical Background

The parasympathetic fibers of the vagal nerve stimulating specifically the stomach arise from the T6 root of the spinal cord. These fibers mainly innervate the gastric body and fundus [18].

Considering these 4 issues, we initially hypothesized that, based on the creation of a somato-autonomic reflex, the stimulation of sensory nerve terminals located in dermatome T6 may cause a reflex, whose efferent pathways end in vagal nerve branches stimulating the gastric wall, similarly to the gastric pacemaker.

PENS OF DERMATOME T6 METHODOLOGY

Urgent PC 200 Neuromodulation System® (Uroplasty, Minnetonka, MN, USA) was used, a device that was originally developed to treat fecal and urinary incontinence. The participants underwent one 30-min session every week for 12 consecutive weeks. Each patient was placed in a supine position without anesthesia, and PENS was delivered by a needle electrode inserted in the left upper quadrant along the medioclavicular line, 2 cm below the ribcage at a 90° angle towards the abdominal wall at a depth of approximately 0.5-1 cm. Successful placement was confirmed by the feeling of electric sensation movement at least 5 cm beyond the dermatome territory. PENS was undertaken at frequency of 20 Hz at the highest amplify (0-20 mA) without causing pain.

RESULTS

An initial study was performed by our group in 2013 [19], comparing 45 morbidly obese patients undergoing PENS of dermatome T6 associated to a 1200 Kcal/day diet, with 45 patients following only the 1200 Kcal/day diet. Trying to eliminate a possible placebo effect associated with the intervention, 15 obese patients undergoing PENS of posterior tibial nerve for fecal incontinence and following the same 1200 Kcal/day diet were also included.

The patients undergoing PENS of dermatome T6 associated with the hypocaloric diet presented a mean weight loss of 7.1 Kg and an excess weight loss of 10.7% vs 2 Kg weight loss and 3.2% excess weight loss in the patients following only the diet. Referring to the appetite perception, as measured by Visual Analogic Scale (ranging from 0 to 10), the patients undergoing PENS of T6 experimented an appetite reduction from 6 to 1.5, while there were no differences in the patients following only the diet and those ones associating the diet with PENS of posterior tibial nerve, demonstrating that the appetite reduction was not just a placebo effect related with the intervention. Dietary compliance after 12 weeks was 93.3% in patients undergoing PENS of dermatome T6, while 50% of the patients following only the diet recognize that they have abandoned it, as the hunger perception was unbearable.

The median treatment week, in which decreased appetite was reported by the patients, was the second week (range 1st -6th).

From these preliminary results, we observed that appetite was associated with diet compliance, which, logically, was related with weight loss. We did not analyze the isolated effect of PENS of dermatome T6 without diet, but this therapy itself does not justify a relevant weight loss. As shown in this study, its main effect is appetite reduction; all of the patients presented with mild feelings of hunger or even absence of hunger after PENS of dermatome T6. Though some weight loss could be obtained just based on less food intake, secondary to the appetite reduction, a specific low calorie diet must be established in order to obtain maximum benefit from the appetite reduction.

In terms of bariatric surgery aims, an excess weight loss over 50% with a final BMI < 35 Kg/m^2 is considered a satisfactory result of a bariatric technique [20]. Only one morbidly obese patient, undergoing PENS of dermatome T6 achieved these goals, and decided to abandon the bariatric surgery program. In these patients, the therapy was planned to achieve a weight reduction before bariatric surgery to reduce the surgical risk. Therefore, after finishing the treatment with PENS of dermatome T6, the patients underwent a bariatric technique. Given these results, PENS of dermatome T6 could not be considered a bariatric approach, as a mean weight loss of 7.1 Kg and a mean excess weight loss of 10.7% are not enough for achieving the success aims. From the results of this first study, it remained unknown whether prolonging the therapy would have added some additional effects and how long the effect of this therapy would have lasted. In the PENS of posterior tibial nerve for treating fecal or urinary incontinence, a secondary treatment period every 2 weeks over a 3-month period has been determined to add some benefits to the first treatment [21, 22].

LONG-TERM EFFECT OF PENS OF DERMATOME T6

All of the patients undergoing PENS of dermatome T6 in this first study presented with BMIs > 35 Kg/m^2, and, as already mentioned, the weight reduction obtained in these patients was not enough to alleviate their morbid obesity. However, in patients presenting with mild obesity or even overweight, this therapy would most likely help them to lose their weight excess, returning to a normal weight status. Thus, looking for an answer for all these questions, we performed a prospective study including 150 consecutive obese patients with BMI between 30 and 40 Kg/m^2, and previous dietary treatment failure [23]. A significant reduction of weight and BMI was observed between pre-treatment values and values after 12 weeks of treatment. This significant reduction was maintained during the long-term follow-up (3 and 9 months after finishing the therapy). Mean weight loss after the 12 weeks of treatment was 11.8 Kg, 3 months after having finished the treatment mean weight loss was 14.6 Kg and 9 months after having finished it 14.5 Kg. Excess weight lost was 66%. 9 months after finishing the treatment (1 year after beginning it), 42% of all the patients presented a BMI within the normal range (BMI: 20 - 25 Kg/m^2) and the rest of the patients were within the range of overweight (BMI: 25 - 30 Kg/m^2).

A significant appetite reduction was observed after finishing the 12 weeks of therapy and maintained during the first 3 months after finishing it, but later on the appetite sensation was gradually restored. 96% of the patients presented a decrease in their appetite to some degree. Dietary compliance after 12 weeks of treatment was 90%, 3 months after finishing the treatment the compliance was 84% and 9 months after finishing the compliance was still 62%.

It is widely known that the main reason for dietary treatment failure is the weight regain once the diet is abandoned. This phenomenon is called the "yoyo effect" [1, 4]. The T6 method allows excellent diet compliance during the therapy, resulting in a mean weight loss of 11.8 Kg after 12 weeks. But the main advantage of this technique is that, after finishing the therapy, the satiating effect lasts for at least 3 months more, allowing a greater weight loss during this period. Surprisingly, 12 months after the beginning of the treatment (9 months after finishing it), appetite is restored, but a weight regain has not been observed. Here we want to remark one fact: the prescribed diet associated to the T6 method is based on common foods eaten by the patients, but with a caloric restriction achieved by reducing the amount of food and limiting hypercaloric elements. Actually, many diets are based on dysbalances between carbohydrates, proteins and fats (e.g., hyperproteic diets) or in the ingestion of commercial nutritional formulas, some of them even mimicking "normal food". The main drawback of these dysbalanced or artificial diest is that the patients often regain the weight lost, once they abandon them and return to their normal and balanced diet [24, 25]. In our opinion, apart from the satiating effect obtained with the PENS of dermatome T6, which is essential for the initial dietary compliance, the key point to achieve a significant weight loss consists in an intense surveillance of the diet by a dietician, adapting the diet to the individual features of each subject and advising some tricks to ease the dietary compliance. Patients can tire of following a low-calorie diet for long time periods, even in absence of hunger sensation, and the role of the dietitian is essential to maintain the motivation. However, the close surveillance by the dietician cannot be prolonged indefinitely. Then, the key point to maintain the weight loss, once the satiating effect of the therapy ends and the close surveillance by the dietitian is also finished, is that the patient has learned during the period of surveillance how to keep a healthy diet and how to adapt it to the personal conditions.

HORMONAL EFFECT OF PENS OF DERMATOME T6

We have already presented the clinical results of dermatome T6, but the neurohormonal effect of this therapy remains still partially unknown. The initial hypothesis postulated that, based on the creation of a somato-autonomic reflex, the stimulation of sensory nerve terminals located in dermatome T6 might cause a reflex, for which the efferent pathways would end in vagal nerve branches stimulating the gastric wall [19]. Similarly to the effect of the gastric pacemaker, once stimulated the stomach, the gastric emptying would be impaired, leading to early satiety and reduced food intake. Moreover, the induction of gastric distension in the fasting state could result in the activation of stretch receptors, causing satiety [8].

Ghrelin is one of the main hormones involved in this satiety mechanism. When the gastric emptying is delayed, stretch receptors located in the gastric fundus inhibit the segregation of ghrelin to the bloodstream. Ghrelin is an orexygen hormone acting in the appetite center in the hypothalamus, increasing the appetite sensation. In cases of decreased ghrelin segregation, the appetite sensation is also reduced, being this the mechanism explaining weight loss and appetite reduction after PENS of dermatome T6 [10, 14-16].

Our group has also performed a clinical trial [26], including 200 obese patients, randomized into 4 groups: Patients undergoing PENS T6 in conjunction with the implementation of a hypocaloric diet (1200 kcal/d) (group 1), patients undergoing PENS T6 and following a normocaloric diet (2000 kcal/d) (group 2), patients undergoing TENS of dermatomes in right iliac fossa (dermatomes T11-T12) and following a hypocaloric diet (1200 kcal/d) (group 3) and those patients following only a 1200 kcal diet (group 4).

The inclusion of group 2 was focused on determining the isolated effect of PENS T6 on weight loss, without association of a hypocaloric diet. Group 3 was included to avoid the placebo effect associated to the intervention as a bias in the study; TENS was used instead of PENS as the local ethics committee did not allow the performance of an invasive procedure for a sham stimulation. Group 4 was included to evaluate the effect obtained just with the hypocaloric diet.

A first blood sample was obtained before beginning the treatment, a second one after finishing the first sesión of neurostimulation (PENS or TENS), a third one before undergoing the last neurostimulation session, a fourth one after finishing this last stimulation and, finally, a fifth one 1 month after the end of the therapy. We wanted to evaluate:

- first sample: baseline levels

- second sample: immediate effect after stimulation (peak)
- third sample: maintained effect after 1 week without treatment, before the 12th stimulation (valley)
- fourth sample: final accumulated effect
- fifth sample: residual effect 1 month after finishing the therapy

In group 4, as there was no stimulation, only blood samples 1, 4, and 5 were obtained. All the blood samples were obtained in fasting status.

Analyzed laboratory data include ghrelin, growth hormone (GH), insulin-like growth factor 1 (IGF-1), glucose, glycated hemoglobin, insulin, triglycerids, total cholesterol, and high density lipoprotein (cholesterol). Insulin resistance was determined by the homeostatic model assessment formula (Insulin x Glucose/405).

50 patients were included in each group. Comparing preinterventional and postinterventional values, only the patients in group 1 (PENS T6 combined with diet) experienced significant reductions of weight and BMI. There were no significant changes of weight and BMI in the other groups.

All of the patients in groups 1 and 2 experienced appetite reduction compared with 30% of the patients in group 3 (TENS T11 combined with diet) and 4% of the patients in group 4. Moreover, 50% of the patients in groups 3 and 70% in group 4 reported an increase in appetite, leading them to abandon the diet in most cases while under study protocol (24% of drop outs in group 3 and 36% in group 4). They reported that this appetite increase was related to a lower caloric intake, while following the prescribed diet. These patients probably lose weight so long they followed the diet, but when they abandoned it, they possibly regained part of the weight lost. Patients undergoing PENS T6, associated or not to a hypocaloric diet, presented significantly higher diet compliance than those ones undergoing TENS of dermatomes T11-T12, and also greater than the patients following only the hypocaloric diet.

Similarly to the immediate effect obtained after finishing the treatments, pairwise analyses revealed a significantly greater excess weight loss in patients following PENS T6 combined with diet, when compared with the other 3 groups. There were no significant differences in excess weight loss between the other 3 groups.

Regarding the evolution of the appetite in each group, there were no significant differences in the appetite quantification between just after finishing the treatment and 1 month later, in groups 1 and 2. However, appetite increased significantly after this period in patients of group 3. Appetite sensation in group 4 remained similar or even slightly lower, as most patients abandoned the diet.

Referring to diet compliance, this remained similar in groups 1 and 2, but decreased significantly in groups 3 and 4.

In the posttreatment values, significant differences could be observed in Ghrelin, GH, IGF-1, glucose, and triglycerids values. Pairwise analysis revealed that ghrelin levels were significantly lower in patients undergoing PENS T6, associated or not to hypocaloric diet. Similarly, GH and IGF-1 values were also significantly lower in the groups 1 and 2, when compared with the other groups. Glucose levels were significantly lower in group 1, when compared with the other groups, without differences between groups 2, 3, and 4. In the same way, triglycerids were significantly lower in group 1, when compared with the other groups, without differences between groups, 2, 3, and 4.

One month after finishing the treatment, biochemical data remain similar to that obtained just after finishing it. The only exception was that glycated hemoglobin and homeostatic model assessment were significantly lower in group 1, when compared with the other groups. These significant differences were obtained 1 month after finishing the treatment, but were not evident in the posttreatment values.

Ghrelin Monitoring

After the first stimulation episode, in patients of group 1 and 2, significant changes in ghrelin levels could not be observed. Curiously, instead of a tendency to decrease just after the first stimulation, ghrelin levels tend to a slight increase. Before the last stimulation (sample 3), ghrelin values were significantly lower than baseline ones in both groups. Just after the last stimulation (sample 4), ghrelin values continue significantly lower than baseline ones, but slightly higher than previous ones. Finally, 1 month after having finished the treatment (sample 5), ghrelin levels remain significantly lower than baseline ones in both groups.

In the patients undergoing PENS T6 (including groups 1 and 2), a direct correlation could be observed between the appetite reduction and the ghrelin reduction after finishing the treatment and 1 month later. Ghrelin reduction in groups 1 and 2 also show a direct correlation with GH reduction after finishing the treatment and 1 month later. Similar correlation could be observed between ghrelin and IGF-1 reduction after finishing the treatment and 1 month later. Correlations between ghrelin levels and other clinical or biochemical parameters could not be demonstrated.

In this study we have demonstrated that PENS T6 induces a reduction in serum ghrelin levels, which was maintained at least up to 1 month after finishing the stimulation. Consequently, a parallel reduction of GH and IGF-1 (stimulated by GH secretion) was also observed. The effect of ghrelin inhibition was not immediate after the first PENS T6. In the monitoring of the plasma levels of ghrelin performed in this study, a significant reduction could not be observed just after finishing the therapy, but a slight nonsignificant increase, both after the first and the last stimulation episodes. The patients did not refer a greater appetite after the stimulation. Thus, the significance of this paradoxical ghrelin augmentation remains unclear. We did not obtain a blood sample just before the second stimulation (7 days after the first one) in order to know if a significant reduction in ghrelin levels was already obtained. However, as described in our previous study, the anorexygenic effect of PENS T6 does not appear at least up to 2 weeks after beginning the therapy. As the reduction of ghrelin between the baseline values and the post-treatment ones correlates with the appetite decrease, we can assume that the decrease in ghrelin levels might appear around the moment of appetite reduction, but further studies must be conducted to confirm this hypothesis.

REFERENCES

[1] Bray, GA. Medical consequences of obesity. *J Clin Endocrinol Metab*, 2004, 2583-2589.
[2] Vest, AR; Heneghan, HM; Agarwal, S; et al. Bariatric surgery and cardiovascular outcomes: a systematic review. *Heart.*, 2012, 98, 1763-1777.
[3] Nguyen, NT; Magno, CP; Lane, KT; et al. Association of hypertension, diabetes, dyslipidemia and metabolic syndrome with obesity: findings from the National Health and Nutrition Examination Survey 1999 to 2004. *J Am Coll Surg*, 2008, 207, 928-934.
[4] Martin, Duce A; Diez, del Val I. Cirugía de la obesidad mórbida. Guías Clínicas de la Asociación Española de Cirujanos. *Madrid, Aran*, 2007.
[5] Sullivan, PW; Ghushchyan, VH; Ben-Joseph, R. The impact of obesity on diabetes, hyperlipidemia and hypertension in the United States. *Qual Life Res*, 2008, 17, 1063-1071.
[6] Kaul, A; Sharma, J. Impact of bariatric surgery on comorbidities. *Surg Clin North Am.*, 2011, 91, 1295-1312.

[7] Buchwald, H; Avidor, Y; Braunwald, E; et al. Bariatric surgery: a systematic review and meta-analysis. *JAMA*, 2004, 292, 1724-1737.

[8] Chen, J. Mechanisms of action of the implantable gastric stimulator for obesity. *Obes Surg*, 2004, 28-32.

[9] Yao, SK; Ke, MY; Wang, ZF; et al. Visceral response to acute retrograde gastric electrical stimulation in healthy human. *World J Gastroenterol*, 2005, 11, 4541-4546.

[10] De Luca, M; Segato, G; Busetto, L; et al. Progress in implantable gastric stimulation: summary of results of the European multi-center study. *Obes Surg*, 2004, 14, 33-39.

[11] Van der Pal, F; Van Balken, MR; Heesakkers, JP; et al. Percutaneous tibial nerve stimulation in the treatment of overactive bladder syndrome: is maintenance treatment a necessity? *BJU Int*, 2006, 97, 547-550.

[12] Boyle, DJ; Prosser, K; Allison, ME; et al. Percutaneous tibial nerve stimulation for the treatment of urge fecal incontinence. *Dis Colon Rectum*, 2010, 53, 432-437.

[13] Pereira, E; Foster, A. Appetite suppression and weight loss incidental to spinal cord stimulation for pain relief. *Obes Surg*, 2007, 17, 1272-1274.

[14] Wang, J; Song, J; Hou, X; et al. Effects of cutaneous gastric electrical stimulation on gastric emptying and postprandial satiety and fullness in lean and obese subjects. *J Clin Gastroenterol*, 2010, 44, 335-339.

[15] Yin, J; Ouyang, H; Wang, Z; et al. Cutaneous gastric electrical stimulation alters gastric motility in dogs: New option for gastric electrical stimulation? *J Gastroenterol Hepatol*, 2009, 24, 149-154.

[16] Abell, TL; Minocha, A; Abidi, N. Looking to the future: electrical stimulation for obesity. *Am J Med Sci*, 2006, 331, 226-232.

[17] Vitton, V; Damon, H; Roman, S; et al. Transcutaneous posterior tibial nerve stimulation for fecal incontinence in inflammatory

bowel disease patients: a therapeutic option? *Inflamm Bowel Dis*, 2009, 15, 402-405.
[18] Moore, KL; Dalley, II AF. *Clinically oriented anatomy.* Philadelphia, Lippincott Wilkins & Williams, 2006, pp. 321-325.
[19] Ruiz-Tovar, J; Oller, I; Diez, M; et al. Percutaneous electrical neurostimulation of dermatome T6 for appetite reduction and weight loss in morbidly obese patients. *Obes Surg.*, 2014, 24, 205-211.
[20] Lemanu, DP; Srinivasa, S; Singh, PP; et al. Laparoscopic sleeve gastrectomy: its place in bariatric surgery for the severely obese patient. *N Z Med J.*, 2012, 125, 41-49.
[21] Monga, AK; Tracey, MR; Subbaroyan, J. A systematic review of clinical studies of electrical stimulation for treatment of lower urinary tract dysfunction. *Int Urogynecol J.*, 2012, 23, 993-1005.
[22] Findlay, JM; Maxwell-Armstrong, C. Posterior tibial nerve stimulation and faecal incontinence: a review. *Int J Colorectal Dis.*, 2011, 26, 265-273.
[23] Ruiz-Tovar, J; Llavero, C. Long – term effect of percutaneous electrical neurostimulation of dermatome T6 for appetite reduction and weight loss in obese patients. *J Laparoendosc Adv Surg Tech*, 2016, 26, 212-215.
[24] Langeveld, M; De Vries, JH. The long-term effect of energy restricted diets for treating obesity. *Obesity (Silver Spring)*, 2015, 23, 1529-1538.
[25] Leeds, AR. Formula food-reducing diets: A new evidence-based addition to the weight management tool box. *Nutr Bull*, 2014, 39, 238-246.
[26] Ruiz-Tovar, J; Llavero, C; Smith, W. Percutaneous electrical neurostimulation of dermatome T6 for short-term weight loss in overweight and obese patients: Effect on ghrelin levels, glucose, lipid and hormonal profile. *J Laparoendosc Adv Surg Tech*, 2017, 27, 241-247.

In: Uses of Electrical Stimulation...　　ISBN: 978-1-53615-036-0
Editor: Jaime Ruiz-Tovar　　© 2019 Nova Science Publishers, Inc.

Chapter 4

PERCUTANEOUS ELECTRICAL NEUROSTIMULATION OF DERMATOME T7 FOR THE TREATMENT OF DIABETES MELLITUS

Jaime Ruiz-Tovar[*]*, MD, PhD and Carolina Llavero*
Electrostimulation Unit, Clinica Garcilaso, Madrid, Spain

ABSTRACT

Endocrine pancreatic segregation is regulated by the autonomic nervous system. Parasympathetic system stimulates insulin production by the beta cells and inhibits the adrenergic discharge by the sympathetic nervous system. Percutaneous neurostimulation (PENS) of dermatome T7 generates a somato-autonomic reflex, whose efferent pathway are the vagal branches specifically stimulating the pancreas. This treatment obtains a decrease of glycemia, based on the reduction of insulin peripherical resistance, but without increasing insulin secretion. This approach obtains

---

[*] Corresponding Author Email: jruiztovar@gmail.com.

a further improvement of the diabetic status, than the one associated with weight loss after following a hypocaloric diet.

Keywords: Percutaneous neurostimulation, dermatome T7, diabetes mellitus, Insulin, HOMA

INTRODUCTION

It has been widely demonstrated that obesity is a major risk for the develoment of type 2 diabetes mellitus (T2DM). In the last decades, the prevalence of T2DM has increased simetrically to obesity. Several studies have estimated that the risk of developing T2DM increases a 4.5% with each Kg over the normality range. Thus, in severe or morbidly obese patients, the probability of developing T2DM at any moment in their life is nearly 100% [1-3]. The insulin resistance is the major determinant for the development of T2DM, and is manifested when the pancreatic Beta cell is unable to segregate more insulin to overcome the peripherical resistance to the insulin action [4].

Weight loss has shown to achieve an improvement in the glycemic profile [5].

Pancreatic endocrine secretion is regulated by the autonomous nervous system. Parasympathetic system stimulates the insulin production by the Beta-cells. The exogenous administration of acetylcholine also increases the release of insulin. Moreover, the parasympathetic activation inhibits the adrenergic release from the sympathetic nervous system [6].

A previous study of our group evaluated the effect of percutaneous electrical neurostimulation (PENS) of dermatome T7 on glycemia, insulin release and resistance to the action of insulin, based on the theory of creation of a somato-autonomic reflex, whose afferent pathway are the sensitive branches of dermatome T7 and the efferent pathway the branches of the vagal nerve, specifically stimulating the pancreas [7].

A prospective randomized clinical trial was performed, including 30 patients in each group. Inclusion criteria were patients with BMI ≥ 30 Kg/m^2 and T2DM under treatment with Metformin. Patients were randomized into 2 groups: those ones undergoing PENS of dermatome T7 associated to hypocaloric diet (1200 Kcal/day) and those ones following only the hypocaloric diet. Blood samples were obtained from the patients before beginning the treatment and 7 days after finishing it.

PENS OF DERMATOME T7 METHODOLOGY

Urgent PC 200 Neuromodulation System® (Uroplasty, Minnetonka, MN, USA) was used, a device that was originally developed to treat fecal and urinary incontinence. The participants underwent one 30-min session every week for 12 consecutive weeks. Each patient was placed in a supine position without anesthesia, and PENS was delivered by a needle electrode inserted in the left upper quadrant along the medioclavicular line, 4 cm below the ribcage (territory of dermatome T7) at a 90° angle towards the abdominal wall at a depth of approximately 0.5-1 cm. Successful placement was confirmed by the feeling of electric sensation movement at least 5 cm beyond the dermatome territory. PENS was undertaken at frequency of 20 Hz at the highest amplify (0-20 mA) without causing pain.

RESULTS

The sample consisted in 80% females and 20% males with a mean age of 48.9 ± 13.8 years, without significant differences in the baseline features between groups.

After finishing the treatment, in the patients undergoing PENS T7 associated to diet, a significant decrease in the glycemia (Mean decrease

62,1 mg/dl; IC95% (41,6-82,6); p = 0,024) and in HOMA (Mean decrease 1,37; IC95% (0,63-2,11); p = 0,014) could be observed. Glycated hemoglobin values tend also to decrease, but did not reach the statistical significance. In the patients following only the diet, statistical differences could not be observed in any of the analyzed parameters, before and after intervention.

Weight Loss and Association with Glycemic Profile

Weight loss is greater in the patients undergoing PENS than in those ones following only diet (6,4 ± 1,4 Kg vs 2,5 ± 0,8 Kg; p = 0,08). However, only in the patients following diet, a correlation between weight loss and HOMA decrease could be determined. In the patients undergoing PENS, HOMA decrease was greater than the weight loss obtained, indicating an eventual effect of PENS T7 on glycemic metabolism, independently of the weight loss.

JUSTIFICATION

PENS effect has been widely demonstrated in the neurostimulation of the posterior tibial nerve for the treatment of urinary or fecal incontinence, creating a somato-somatic reflex [8, 9]. This will be widely discussed elsewhere in this book.

Our group was the one who discovered the use of PENS of dermatome T6 for appetite reduction and consequently to obtain a significant weight loss, as it will be also discussed in another chapter [10]. PENS of dermatome T7 is based on the same background than PENS of T6: A somato-autonomic reflex, indirectly stimulating the pancreas.

Weight loss alone has shown to improve the glycemic profile, mainly secondary to the reduction of the insulin resistance in peripheric tissue.

Moreover, weight loss also determines a reduction of the hyperinsulinism derived from obesity [11, 12].

PENS of dermatome T7 obtains a greater reduction of glycemia and insulin resistance, as measured by HOMA. Curiously, instead of showing an increase of insulin release, as the pancreas is stimulated by the parasympathetic autonomous nervous system, the serum insulin levels remained unaltered. Thus, the reduction of glycemia is just secondary to the decrease in the insulin resistance and improvement of the glucose profile in the peripherical tissues. As previously mentioned, weight loss also achieves a reduction of insulin resistance, but in the patients undergoing PENS T7, the reduction of HOMA was significantly greater than that expected for the weight loss.

Parasympathetic stimulation of the pancreas, apart from the direct effects on the gland, inhibits the sympathetic nervous system and consequently decreases the release of contraregulatory hormones (epinephrine, growth hormone and cortisol). These hormones induce an augmentation in the endogenous synthesis of glucose and inhibit its use in the peripherical tissues, mainly in the skeletikal muscle, contributing to the insulin resistance [13]. In our opinion, this is the main physiopathological mechanism to justify the effects of PENS T7 on the glucose homeostasis.

However, other possible mechanisms can be also involved. The dermatomes are not exact segments of skin, receiving innervation from only one branch of the spinal cord. During PENS of dermatome T7, collateral branches of the nerves might also stimulate partially the reflex of dermatome T6. The patients undergoing PENS T7 alre report a certain reduction of appetite sensation. T6 stimulation produces a delay in the gastric emptying and inhibits the ghrelin release by the gastric fundus [14, 15]. The ghrelin reduction increases the pancreatic segregation of insulin and reduces its peripherical resistance, especially in the adipose tissue. Thus, this action might be more remarkable in obese subjects [16]. The inhibition of ghrelin might also favor a decrease of HOMA.

Future studies must evaluate the long-term benefit of PENS T7 on the homeostasis of glucose, and analyze the effect on epinephrine, growth hormone, cortisol and ghrelin levels, among other hormones.

REFERENCES

[1] Portero McLellan KC, Wyne K, Villagomez ET, Hsueh WA. Therapeutic interventions to reduce the risk of progression from prediabetes to type 2 diabetes mellitus. *Ther Clin Risk Manag.* 2014;10:173-188.

[2] Misra A, Khurana L. Obesity and the metabolic syndrome in developing countries. *J Clin Endocrinol Metab.* 2008;93:S9–S30.

[3] International Diabetes Federation. *IDF Diabetes Atlas.* Brussels, Belgium: International Diabetes Federation; 2013.

[4] Igel LI, Powell AG, Apovian CM, Aronne LJ. Advances in medical therapy for weight loss and the weight-centric management of type 2 diabetes mellitus. *Curr Atheroscler Rep.* 2012;14:60-69.

[5] Castagneto M, Mingrone G. The effect of gastrointestinal surgery on insulin resistance and insulin secretion. *Curr Atheroscler Rep.* 2012;14:624-630.

[6] Begg DP, Woods SC. Interactions between the central nervous system and pancreatic islet secretions: a historical perspective. *Adv Physiol Educ.* 2013; 37: 53–60.

[7] Ruiz-Tovar J, Llavero C, Ortega I, Diez M, Zubiaga L, Calpena R. [Percutaneous electrical neurostimulation of dermatome T7 improves the glycemic profile in obese and patients with Type 2 diabetes mellitus. A randomized clinical trial]. *Cir Esp* 2015;93:460-465.

[8] Van der Pal F, Van Balken MR, Heesakkers JP, Debruyne FM, Bemelmans BL. Percutaneous tibial nerve stimulation in the treatment of overactive bladder syndrome: is maintenance treatment a necessity? *BJU Int* 2006;97:547-550.

[9] Boyle DJ, Prosser K, Allison ME, Williams NS, Chan CL. Percutaneous tibial nerve stimulation for the treatment of urge fecal incontinence. *Dis Colon Rectum.* 2010;53:432-437.

[10] Ruiz-Tovar J, Oller I, Diez M, Zubiaga L, Arroyo A, Calpena R. Percutaneous electrical neurostimulation of dermatome T6 for appetite reduction and weight loss in morbidly obese patients. *Obes Surg* 2014;24:205-211.

[11] Wing RR, Blair EH, Bononi P, Marcus MD, Watanabe R, Bergman RN. Caloric restriction per se is a significant factor in improvements in glycemic control and insulin sensitivity during weight loss in NIDDM patients. *Diabetes Care* 1994; 17:30-36.

[12] Goldstein DJ. Beneficial health effects of modest weight loss. *Int J Obes Relat Metab Disord* 1992;16:397-415.

[13] Lucidi P, Rossetti P, Porcellati F, Pampanelli S, Candeloro P, Andreoli AM, et al. Mechanisms of insulin resistance after insulin-induced hypoglycemia in humans: the role of lipolysis. *Diabetes* 2010; 59: 1349–1357.

[14] Chen J. Mechanisms of action of the implantable gastric stimulator for obesity. *Obes Surg* 2004;28-32.

[15] De Luca M, Segato G, Busetto L, Favretti F, Aigner F, Weiss H, et al. Progress in implantable gastric stimulation: Summary of results of the European Multi-center study. *Obes Surg* 2004;14:33-39.

[16] Dezaki K. Ghrelin function in insulin release and glucose metabolism. *Endocr Dev.* 2013;25:135-143.

In: Uses of Electrical Stimulation... ISBN: 978-1-53615-036-0
Editor: Jaime Ruiz-Tovar © 2019 Nova Science Publishers, Inc.

Chapter 5

POSTOPERATIVE PERCUTANEOUS ABDOMINAL ELECTRICAL STIMULATION (PPAES) FOR ILEUS IN PATIENTS UNDERGOING COLORECTAL RESECTION

Pedro Moya[1,*], *Manuel Ferrer-Márquez*[1], *Francisco Rubio-Gil*[1], *Rafael Calpena*[2] *and Angel Reina*[1]

[1]Department of General Surgery, Division of Colorectal Surgery, University Hospital of Torrecardenas, Almeria, Spain
[2]Department of Surgery and Pathology, Miguel Hernández University, Elche, Spain

ABSTRACT

Background. Despite progress in recent years in the surgical management of patients with colorectal cancer, such as Enhanced Recovery After Surgery programmes, postoperative ileus is still frequent.

---

[*] Corresponding Author Email: pedromoyaforcen@gmail.com.

Transcutaneous electrical nerve stimulation is a non-pharmacological intervention strategy for pain relief. It has minimal side effects, and, in theory, it may have beneficial effects on postoperative ileus. Objective. To determine postoperative percutaneous abdominal electrical stimulation reduces postoperative ileus in patients undergoing colorectal resection. Design Settings. Prospective randomised trial with two parallel treatment groups: control (no intervention) or stimulation (stimulation for 3 days postoperatively). Patients. Patients who underwent colorectal resection. Main Outcomes Measures. The time to first stool and flatulence, the duration of the total hospital stay and the appearance of ileus were recorded. Results. A total of 42 patients were randomised. At baseline, both groups were comparable in regards to age, sex, surgical risk, comorbidities, and analytical and nutritional parameters. The median length of postoperative hospital stay was 5 days for all patients (3-13 days). The median length of postoperative hospital stay was six days in the control group (4-13) and 4 days in the stimulation group (3-6), (p = 0.000). Overall, the average time to first flatulence and first deposition were 1.73 days (0-7) and 3.38 days (0-8), respectively. These times were significantly different between the control (2.85 and 4.75 days) and stimulation groups (0.60 and 2 days) (p = 0.000 and p = 0.000, respectively). Limitations. The number of cases was too small to establish robust conclusions. Additionally, despite including two comparable groups, we included various surgical procedures. Conclusions. The use of transcutaneous electrical nerve stimulation reduced postoperative ileus after colorectal surgery and stimulated bowel motility.

Keywords: postoperative ileus, transcutaneous electrical nerve stimulation, colorectal surgery

INTRODUCTION

Postoperative ileus (POI) is a common complication after colorectal surgery, and the incidence is greater than 10% [1, 2]. POI is associated with increased morbidity and a prolonged hospital stay [3-6]. Clinically, POI presents as intolerance to oral feeding, distended abdomen, nausea and/or vomiting, diminished or absent bowel sounds and failure to pass flatus and/or bowel movements [7].

Although perioperative care can be improved through several measures, such as encouraging early mobilisation and performing minimally invasive surgery and standard anaesthetic protocols (as part of the Enhanced Recovery After Surgery (ERAS) guidelines), POI cannot be completely avoided, and the incidence remains high [8].

PATHOPHYSIOLOGY OF POSTOPERATIVE ILEUS

Postoperative ileus is complex and multifactorial, and neurogenic mechanisms appear to play the most important role in early POI. Sympathetic stimulation inhibits gastrointestinal motility, whereas parasympathetic activity primarily stimulates motility. After surgery, the sympathetic system tends to be substantially more active than the parasympathetic system leading to decreased motility and causing POI [9].

TRANSCUTANEOUS ELECTRICAL NERVE STIMULATION

Transcutaneous electrical nerve stimulation (TENS) is a non-pharmacological intervention for pain relief. TENS has minimal side effects [10], and, in theory, it may have beneficial effects on intestinal peristalsis by activation of the parasympathetic system, through the stimulation of the vagus nerve, when electrical stimulation is done at the abdominal level.

We determined whether postoperative percutaneous abdominal electrical stimulation (PPAES) reduces the incidence of POI among patients undergoing colorectal resection.

METHODS

Participants

Patients preoperatively diagnosed with colorectal cancer were randomised in two parallel treatment groups: control (no intervention) or stimulation (PPAES for 3 days postoperatively).

Inclusion and Exclusion Criteria

Inclusion Criteria

All included patients were ≥18 years of age and were scheduled to undergo colorectal cancer surgery without the need for any further surgical procedures. The patients signed a written consent form.

Exclusion Criteria

All patients who did not meet the inclusion criteria were excluded. The other exclusion criteria included the need for emergency surgery, an American Society of Anaesthesiologists (ASA) class of IV, the inability to provide informed consent, psychiatric disorders, HIV, pregnancy/breastfeeding, bowel obstruction, or uncontrolled infection.

Postoperative Percutaneous Abdominal Electrical Stimulation

Standard TENS was performed. For the PPAES group, subjects underwent one 30 min session for three days postoperatively. Subjects were placed in the supine position. The patients' periumbilical areas were cleaned with 75% alcohol swabs and a self-adhesive electrode was placed. Without anaesthesia, a needle electrode was inserted 4-5 cm lateral to the self-adhesive electrode to a depth of approximately 0.5-1 cm. The stimulator delivered a current of 1.5 mA at a fixed pulse frequency of 20 Hz and a pulse width of 200 μseconds. The pulses had a square waveform and a resistance of 500 to 4000 ohms. When the treatment time was complete, the self-adhesive and needle electrodes were removed.

Outcome Measures

Patient baseline characteristics at the time of surgery (age, gender, and the American Society of Anaesthesiologists (ASA) score) and major comorbidities were recorded for each patient.

The time to first stool and flatulence were recorded. The duration of the total hospital stay was also recorded. The appearance of ileus, defined as the presence of abdominal distension, lack of intestinal transit and requirement of a nasogastric tube, was scored. The 30-day postoperative complications were recorded. Complications were defined as any deviation from the normal postoperative course and were divided into minor and major complications. Minor complications included minor risk events, such as wounds infections (occurred at the bedside), urinary tract infections or POI. Major complications included potentially life-threatening complications and complications (e.g., anastomotic leaks, abdominal abscesses, or pneumonia) requiring surgical, endoscopic or radiological interventions. The duration of hospital stay and rates and causes of readmission were also documented.

Statistical Analyses

Statistical analyses between the 2 groups were performed using SPSS version 22 (SPSS Inc. Chicago, IL, US). The data were analysed with the intention to treat principle. The data are presented as the means ± standard deviations or as medians and interquartile ranges as appropriate. For dichotomous outcomes, the means of the treatment groups were compared with the χ^2 test. The Mann–Whitney U and Kruskal–Wallis tests were used for continuous, not normally distributed outcomes. For continuous normally distributed data, the ANOVA test was used. Univariate and multiple linear or logistic regression analyses were performed to determine the effect of stimulation.

RESULTS

Patients

A total of 42 patients were randomised. In total, 2 patients opted not to participate, and 40 patients completed the 3-day postoperative phase of the study. The median patient age was 65 years (45-81); and 50% were women (20 patients). The patients were distributed by ASA classification I, II, and III as 45% (18), 30% (12) and 25% (10), respectively. At baseline, the two groups were comparable according to age, sex, surgical risk, and comorbidities (Table 1).

Patient Compliance and Tolerability of Transabdominal Stimulation

All patients in the stimulation group completed the postoperative stimulation treatment. No complications due to PPAES were observed.

Table 1. Characteristics and Surgical Procedures of the Two Groups

Patients characteristics	Without stimulation N = 20	With stimulation N = 20	p
Age (years)	64 (48-75)	66 (45-81)	0.458
Gender – Female – Male	8 (40%) 12 (60%)	12 (60%) 8 (40%)	0.206
Surgical risk – ASA I – ASA II – ASA III	9 (45%) 5 (25%) 6 (30%)	9 (45%) 7 (35%) 4 (20%)	0.693
Surgical approach – Laparoscopy – Conventional	17 (85%) 3 (15%)	16 (80%) 4 (20%)	0.677
Surgical procedure – Right hemicolectomy – Left hemicolectomy – Sigmoidectomy – Anterior resection – Subtotal colectomy	6 (30%) 1 (5%) 6 (30%) 6 (30%) 1 (5%)	7 (35%) 1 (5%) 7 (35%) 4 (20%) 1 (5%)	0.968

Quantitative variables are expressed as medians plus minimum and maximum values; qualitative variables are expressed as absolute numbers and percentages.

Surgery

In total, 82.5% (33) of the patients underwent laparoscopic surgery and 17.5% (7) underwent open surgery. None of the laparoscopically intervened patients required conversion to laparotomy. Sigmoidectomy and right hemicolectomy comprised the majority of procedures performed (65%). There was no significant difference between the two groups regarding the operative time and estimated intraoperative blood loss. Table 1 presents the surgical techniques and procedures according to each group.

Postoperative Hospital Stay, Gastrointestinal Transit and Postoperative Morbidity/Mortality

The median length of the postoperative hospital stay was 5 days (3-13 days): six days in the control group (4-13) and 4 days in PPAES group (3-6), (p = 0.000). Overall, the average times to first flatulence and first deposition were 1.73 days (0-7) and 3.38 days (0-8). The time to first flatulence and first deposition were 2.85 and 4.75 days in the control group and 0.60 and 2 days in the PPAES group, respectively (p = 0.000 and p = 0.000).

Table 2. Postoperative hospital stay, Gastrointestinal transit and Postoperative morbidity/mortality

Outcome variable	Without stimulation N = 20	With stimulation N = 20	Difference	95% CI for difference	P
First passing of flatus[a]	2,85 ± 2,277	0,60 ± 0,754	2,250	1,141-3,359	0,000
First defaecation[a]	4,75 ±1,773	2,00 ± 1,654	2,750	1,652-3,848	0,000
Complication rate[b]	35%	15%	-	RR 0,238 (0,071-1,518)	0,144
Wound infection	10%	5%	-	RR 0,474 (0,039-5,688)	0,548
Ileus	30%	0%	-	RR 0,412 (0,276-0,615)	0,008
Lower intestinal bleeding	5%	10%	-	RR 2,111(0,176-25,349)	0,548
Hospital stay[a]	7,10 ± 2,713	4,53 ± 0,905	2,574	1,247-3,900	0,000

[a]Time in days.
[b]Note that a patient may experience more than one complication.
RR Relative risk.

Approximately 75% of patients had an uneventful postoperative course without complications. As shown in Table 2, patients who received PPAES experienced fewer complications (15% vs. 35%, p = 0.144). The most common surgical complication was POI, which occurred in six cases in the control group and none in the PPAES group (p = 0,008).

In addition, 2 patients in the control group and one in the PPAES group developed surgical site infections. One patient in the control group and two in the PPAES group developed lower gastrointestinal bleeding. No other complications were observed.

Discussion

Despite advances in surgical techniques and improvements in postoperative care, POI remains a very important clinical determinant of short-term morbidity and mortality following colorectal surgery. In this trial, postoperative transabdominal stimulation reduced the incidence of POI. Patients receiving PPAES postoperatively experienced a quicker recovery of intestinal transit compared to patients who received the traditional care.

Normal bowel function is a complex interaction between GI motility, mucosal transport and defecation reflexes. POI may occur after every abdominal surgical procedure [13]. POI presents clinically as the inability to tolerate food, absence of bowel sounds, lack of flatus and defecation and the presence of abdominal distention. Gastrointestinal motility is expected to return to normal within 2–3 days postsurgery [11]. When motility does not occur, POI should be suspected.

Neurogenic mechanisms appear to play the most important role in early POI. Sympathetic stimulation inhibits gastrointestinal motility, whereas parasympathetic activity primarily stimulates motility. After surgery, the sympathetic system tends to be substantially more active than the parasympathetic system leading to decreased motility and causing POI [11]. Nevertheless, other mechanisms also contribute to prolonged POI. Subcutaneous stimulation may activate the parasympathetic system. This parasympathetic activation produces an early recovery of gastrointestinal transit and reduces the incidence of ileus during the postoperative period.

The onset of POI is considered to be multifactorial, and a variety of preventative methods have been investigated. The use of a nasogastric tube has traditionally been used to decrease the incidence of POI, although subsequent studies have shown their inefficiency [14]. Chewing gum is hypothesised to reduce POI by stimulating early recovery of gastrointestinal function through cephalo-vagal stimulation [15]. However, only a small benefit in reducing time to flatus and time to bowel motions following abdominal surgery has been described, and the benefits have a limited clinical significance and almost always occur outside of ERAS programmes [14]. Multiple pharmacologic agents have been tested in attempts to reduce or eliminate the incidence of POI [16]. β blockers of adrenergic receptors (neostigmine), acetylcholine inhibitors (cisapride), and serotonin receptor antagonists (metoclopramide) have been used with varying effects. These agents are also accompanied by multiple side effects, including abdominal cramping and significant cardiovascular events; thus, their use limited [16]. More recently, alvimopan, an elective mu opioid receptor, has been associated with faster return of bowel function, lower incidence of POI, shorter hospitalisation, and a significant cost savings, especially during open surgery (no benefit occurred when performed with laparoscopic surgery) [17]. PPAES is a safe, inexpensive procedure and produces an early restoration of normal gastric transit after surgery, whether it is used after open or laparoscopic surgery. PPAES is a simple, inexpensive way to reduce POI without side effects.

Our study had some limitations. The number of cases was too small to establish robust conclusions. Additionally, despite including two comparable groups, we included various surgical procedures and two paths (laparoscopic and conventional).

The mechanism of action of electrical stimulation and whether its success is due to the activation of the parasympathetic system, which is inhibited during the early postoperative period, or another mechanism, remains to be determined. Another issue complicating the interpretation of these results is the role of pacing patients whose ileus movement is already established (whether related to surgery or other causes).

CONCLUSION

Based on our data from the present randomised study, postoperative PPAES is a simple, inexpensive tool that stimulates bowel motility and reduces POI after colorectal surgery. PPAES is associated with a faster return of bowel function, a lower incidence of POI and a shorter period of hospitalisation (and thus significant cost savings).

REFERENCES

[1] Moghadamyeghaneh Z, Hwang GS, Hanna MH, Phelan M, Carmichael JC, Mills S, et al. Risk factors for prolonged ileus following colon surgery. *Surg Endosc 2015*; Epub ahead of print.

[2] Wolthuis AM, Bislenghi G, Fieuws S, de Buck van Overstraeten A, Boeckxstaens G, D'Hoore A. Incidence of prolonged postoperative ileus after colorectal surgery: a systematic review and meta-analysis. *Colorectal Dis 2015*; Epub ahead of print.

[3] Delaney CP, Senagore AJ, Viscusi ER, Wolff BG, Fort J, Du W, et al. Postoperative upper and lower gastrointestinal recovery and gastrointestinal morbidity in patients undergoing bowel resection: pooled analysis of placebo data from 3 randomized controlled trials. *Am J Surg* 2006; 191: 315-319.

[4] Luckey A, Livingston E, Tache Y. Mechanisms and treatment of postoperative ileus. *Arch Surg* 2003; 138: 206-214.

[5] Mattei P, Rombeau JL. Review of the pathophysiology and management of postoperative ileus. *World J Surg* 2006; 30: 1382-1391.

[6] Baig MK, Wexner SD. Postoperative ileus: a review. *Dis Colon Rectum* 2004; 47: 516-526.

[7] Schwenk W, Böhm B, Haase O, Junghans T, Müller JM. Laparoscopic versus conventional colorectal resection: a prospective randomised study of postoperative ileus and early postoperative feeding. *Langenbecks Arch Surg* 1998; 383: 49–55.

[8] Peters EG, Smeets BJ Dekkers M, Buise MD, de Jonge WJ, Slooter GD, et al. The effects of stimulation of the autonomic nervous system via perioperative nutrition on postoperative ileus and anastomotic leakage following colorectal surgery (SANICS II trial): a study protocol for a double-blind randomized controlled trial. *Trials* 2015; 27: 20.

[9] Zeinali F, Stulberg JJ, Delaney CP. Pharmacological management of postoperative ileus. *Can J Surg.* 2009; 52: 153–157.

[10] Bjerså K, Jildenstaal P, Jakobsson J, Egardt M, Fagevik Olsén M. Adjunct high frequency transcutaneous electric stimulation (TENS) for postoperative pain management during weaning from epidural analgesia following colon surgery: results from a controlled pilot study. *Pain Manag Nurs 2015*; Epub ahead of print.

[11] García-Olmo D, Lima F. Postoperative paralytic ileus. *Cir Espan* 2001; 69: 275-280.

[12] Dauchel J, Schang JC, Kachelhoffer J, Eloy R, Grenier JF. Gastrointestinal myoelectrical activity during the postoperative period in man. *Digestion* 1976; 14: 293-303.

[13] Wolff BG, Pembeton JH, van Heerden JA, Beart RW Jr, Nivatvongs S, et al. Elective colon and rectal surgery without nasogastric decompression. A prospective, randomized trial. *Ann Surg* 1989; 209: 670–675.

[14] Short V, Herbert G, Perry R, Atkinson C, Ness AR, Penfold C. Chewing gum for postoperative recovery of gastrointestinal function. *Cochrane Database Syst Rev.* 2015; 2: CD006506.

[15] Su'a BU, Pollock TT, Lemanu DP, MacCormick AD, Connolly AB, Hill AG. Chewing gum and postoperative ileus in adults: a systematic literature review and meta-analysis. *Int J Surg.* 2015; 14: 49-55.

[16] Keller D, Stein SL. Facilitating return of bowel function after colorectal surgery: alvimopan and gum chewing. *Clin Colon Rectal Surg.* 2013; 26: 186-190.

[17] Adam MA, Lee LM, Kim J, Shenoi M, Mallipeddi M, Aziz H, et al. Alvimopan provides additional improvement in outcomes and cost savings in enhanced recovery colorectal surgery. *Ann Surg.* 2015; Epub ahead of print.

In: Uses of Electrical Stimulation... ISBN: 978-1-53615-036-0
Editor: Jaime Ruiz-Tovar © 2019 Nova Science Publishers, Inc.

Chapter 6

ELECTRICAL STIMULATION FOR THE MANAGEMENT OF POSTOPERATIVE PAIN

Andrés García Marín[1],[*] and Mercedes Pérez López[2]
[1]Department of Pathology and Surgery,
University Miguel Hernández, Elche, Spain
[2]Department of Nursery, University Hospital San Juan de Alicante,
Alicante, Spain

ABSTRACT

Electrical nerve stimulation is a non-pharmacological analgesic method used in chronic pain management, whose efficacy to reduce postoperative pain, vomiting and improve the quality of life in digestive and endocrine surgeries, is being evaluated in clinical trials.

Keywords: pain, visual analogue scale, electrical nerve stimulation, transcutaneous electrical nerve stimulation, electroacupuncture, transcutaneous electrical acupoint stimulation

---

[*] Corresponding Author Email: andres.garciam@goumh.umh.es.

INTRODUCTION

Electrical nerve stimulation (ENS) is a non-pharmacological analgesic method widely used in chronic pain management (head, neck, lower back), whose efficacy in postoperative pain is being evaluated. The reasons for its analgesic effect are not well known but there are many theories: (a) Inhibition of $A\delta$ and C nerve fibers that transmit pain signals in the dorsal horn of the spine. (b) Release of endogenous opiods. Leonard et al. described that the naloxone administration neutralized the analgesic effects of transcutaneous electrical nerve stimulation (TENS). (c) Release of adenosine which produces a local vasodilation and drainage of algiogenic substances resulting from tissue catabolism. Sawynok et al. showed that caffeine consumption reduces analgesic effect of TENS because of it is an antagonist of adenosine receptors [1-5].

A summary of the studies related to ENS and postoperative pain is presented below:

a) Inguinal hernia surgery: DeSantana et al. conducted a randomized, clinical trial comparing 40 patients undergoing unilateral inguinal hernia repair, who received TENS (20 active versus 20 placebo) applied for 30 minutes through 4 electrodes placed around the incision 2 and 4 hours after surgery. They found significant differences in postoperative pain, measured by 10-point numeric rating scale, at 2, 4, 8 and 24 hours after surgery [2]. Eidy et al. conducted a randomized, clinical trial comparing 66 male patients aged 20-50 years with American Society of Anesthesiologists (ASA) classification I-II, undergoing unilateral inguinal hernia repair with Lichtenstein technique, who received TENS 1 hour before surgery (33 active versus 33 placebo). They found significant differences in postoperative pain, measured by visual analogue scale (VAS) at 2 hours (3.54 ± 1.48 versus 5.12 ± 1.41) and 4 hours (4.0 ± 1.5 versus 4.76 ± 1.36) after surgery whereas no significant differences were observed at 6 and 12 hours [1]. Dalamagka et al.

conducted a clinical trial comparing 54 male patients aged under 75 years, ASA I-II, undergoing unilateral inguinal hernia repair with Lichtenstein technique randomized in three groups, I (preoperative, intraoperative and postoperative electroacupuncture [EA]), II (preoperative and postoperative) and III (placebo group with placement of needles but without skin penetration). They found significant differences between groups I-II and III in terms of pain (VAS) and anxiety (State-Trait Anxiety Spielberger Inventory) evaluated at 30 min, 90 min, 10 hours and 24 hours after surgery but there were no differences between groups I and II [6]. Yilmaz et al. conducted a randomized, clinical trial comparing 52 patients undergoing unilateral inguinal hernia repair, who received postoperative TENS (active versus placebo), five times for 30 minutes each. They found that pain and satisfaction scores were significantly better in active TENS group [7].

b) Hemorrhoidectomy: Chiu et al. conducted a randomized, clinical trial comparing 60 patients undergoing hemorrhoidectomy, who received TENS (30 active versus 30 placebo) and they described significant differences in postoperative pain, measured by VAS at 8 (4.1 ± 0.5 versus 5.9 ± 0.5), 12 (3.5 ± 0.4 versus 5.7 ± 0.5), 16 (2.3 ± 0.3 versus 4.1 ± 0.4) and 24 hours (1.9 ± 0.2 versus 3.2 ± 0.4) and morphine use (6.2 ± 1.3 mg versus 11.6 ± 2.2 mg), as well as active TENS group tended to have a lower rate of postoperative acute urinary retention and need for analgesic than control group [8]. Yeh et al. conducted a randomized, clinical trial comparing 80 patients undergoing hemorrhoidectomy, who received transcutaneous acupoint electrical stimulation (TAES), a combination of TENS and traditional Chinese acupoints, in four 20 minutes sessions (39 active versus 41 placebo). They found significant differences in postoperative pain (VAS score) and anxiety (State Anxiety Inventory) [9].

c) Laparoscopic cholecystectomy: Mi et al. conducted a randomized, clinical trial comparing 100 patients ASA I-II, undergoing laparoscopic cholecystectomy, who received TAES (50 active versus 50 placebo). They found a significantly higher global score of quality of recovery-40 questionnaire (QoR-40) and its all domains (emotional state, physical comfort, psychological support, self-care ability and pain) in active TAES at 4 hours, 8 hours, 24 hours and 48 hours after surgery. As well as, the dosages of intraoperative remifentanil and the rates of nausea and vomiting were lower in the active TAES group [10].

d) Appendectomy: Conn et al. conducted a three-arm, randomized, clinical trial comparing 42 patients aged >12 years, undergoing open appendectomy due to an acute appendicitis with or without local peritonitis, who received postoperative TENS (15 active versus 13 placebo) versus 14 patients in control group. They found a significant decrease of pain severity and analgesic intake in TENS patients in comparison with control group. However, there were no differences between active and placebo TENS, so they did not recommend routine TENS [11]. Lee et al. are conducting a three-arm, randomized, clinical trial comparing 138 patients aged 19-65 years undergoing laparoscopic appendectomy due to nonperforated acute appendicitis, who receive EA (46 active versus 46 placebo) versus 46 patients in control group. Their protocol treatment includes four treatments for 30 minutes (1 hour preoperative, 1 hour postoperative, morning and afternoon on the day after surgery) and the primary outcome is the pain intensity in the abdomen measured by the 11-point pain intensity numerical rating scale 24 hours after surgery [12]. Kim et al. are conducting a three-arm, randomized, clinical trial comparing 87 patients aged >20 years, undergoing laparoscopic appendectomy due to an acute appendicitis, who receive EA (29 active versus 29 placebo) versus 29 control group. Their protocol treatment includes four treatments for 30 minutes twice a day from 2 hours after surgery being the primary outcome the time to first passing flatus and secondary outcomes the assessment of postoperative pain measured by VAS and the rate of nausea and vomiting [13].

e) Colorrectal: Bjersa et al. conducted a randomized, clinical trial comparing patients, undergoing open lower abdominal colorrectal surgery, who received postoperative TENS (active versus placebo) during weaning from epidural analgesia. They found significant decrease of pain meausured by VAS in active TENS [14]. Kim et al. are conducting a three-arm, randomized, clinical trial comparing 60 patients, aged 18-80 years, ASA I-II, undergoing elective laparoscopic surgery due to colorrectal cancer, who receive ENS (20 TEAS versus 20 TENS) and control, whose outcomes are postoperative pain, nausea and vomiting, anxiety and quality of life [15].

e) Thyroid surgery: Chen et al. conducted a randomized, clinical trial comparing 84 female patients aged 18-60 years, ASA I-II, undergoing thyroidectomy, who received preoperative TEAS (42 active versus 42 placebo). They found a significantly higher global QoR-40 score and its domains (only in emotional state, physical comfort and pain) in active TEAS at 24 hours after surgery as wells as a lower rate of nausea and vomiting (24.4% versus 61.9%) and dizziness (29.3% versus 64.2%) [16]. Park et al. conducted a randomized, clinical trial comparing 100 female patients aged 20-60 years, ASA I-II, undergoing thyroidectomy, who received intraoperative TENS applied to the trapezius muscle versus control. They found a significant lower posterior neck pain in TENS measured by 11-point numerical rating scale at 30 minutes, 6 hours, 24 hours and 48 hours after surgery [17].

f) Other surgeries: Tokuda et al. conducted a randomized, clinical trial comparing 48 patients, undergoing abdominal surgery, who received postoperative TENS (active versus placebo) and control 1 hour a day for 3 days. They found a significant reduction in postoperative pain and improvement in pulmonary functions (vital capacity, cough peak flow) in active TENS [18]. Sun et al. conducted a randomized, clinical trial comparing 380 patients undergoing laparoscopic surgery, who received TEAS (preoperative, preoperative and intraoperative, preoperative and postoperative versus placebo). They found a significantly lower pain intensity in combined therapy at 6 hours after surgery and a higher rate of patient satisfaction in active TEAS. There were no significant differences in intraoperative anesthetic consumption, postoperative nausea and vomiting and time of the first postoperative flatus between the four groups [19]. Xie et al. conducted a three-arm randomized, clinical trial comparing 60 patients, undergoing esophagectomy, who received intraoperative TEAS (active versus placebo) and control. They found a significant lower postoperative pain (VAS score) at 2, 12, 24 and 48 hours and total postoperative sef-administration of sufentanil in TEAS group [20]. Bjersa et al. conducted a randomized, clinical trial comparing 55 patients, undergoing pancreatic surgery by horizontal incision, who received postoperative TENS (active versus placebo) during the 24 hours of transition from epidural to general analgesia. They found that postoperative pain and analgesic consumption were lower in active TENS group but not statistically significant [21].

TENS has been also compared with transversus abdominis plane (TAP) block, a regional analgesic technique which blocks T6-L1 nerve branches and has a role in postoperative analgesia for infraumbilical surgeries. Chatrath et al. conducted a randomized, clinical trial comparing 60 patients aged 20-65 years, ASA I-II, undergoing infraumbilical surgeries, who received active postoperative TENS (30 patients) for 30 minutes every 2 hours until 24 hours versus TAP block (30 patients). They found that both provided an excellent postoperative analgesia; however, TAP block was a better modality of analgesia due to a significantly lower VAS score in the first 4 hours after surgery without significant differences from 6 hours. The incidence of nausea and vomiting was significantly lower in TENS group whereas the patient satisfaction score was significantly higher in TAP block group at 24 hours [22].

In conclusion, ENS is a good non-pharmacological method of analgesia that has shown a reduction of postoperative pain, nausea and vomiting and anxiety and an increase of patient satisfaction in different digestive and endocrine surgeries in comparison with placebo or standard analgesia.

REFERENCES

[1] Eidy M, Fazel M, Janzamini M, Rezaei MH, Moravveji AR. Preemptive analgesic effects of transcutaneous electrical nerve stimulation (TENS) on postoperative pain : a randomized, double-blind, placebo-controlled trial. *Iran Red Crescent Med J* [Internet]. 2016;48 (4). Available at: https://www.ncbi.nlm.nih.gov/pmc/articles/PMC4893426/pdf/ircmj-18-04-35050.pdf

[2] DeSantana JM, Santana-Filho VJ, Guerra DR, Sluka KA, Gurgel RQ, daSilva WM Jr. Hypoalgesic effect of the transcutaneous electrical nerve stimulation following inguinal herniorrhaphy: a randomized controlled trial. *J Pain*. 2008;9(7):623-629.

[3] Ren W, Tu W, Jiang S, Cheng R, Du Y. Electroacupuncture improves neuropathic pain: adenosine, adenosine 5'triphosphate disodium and their receptors perhaps change simultaneously. *Neural Regen Res*. 2012;7(33):2618-2623.

[4] Sawynok J. Adenosine receptor targets for pain. *Neuroscience*. 2016;338:1-18.

[5] Leonard G, Goffaux P, Marchand S. Deciphering the role of endogenous opioids in high-frequency TENS using low and high doses of naloxone. *Pain.*2010;151(1):215-219.

[6] Dalamagka M, Mavrommatis C, Grosomanidis V, Karakoulas K, Vasilakos D. Postoperative analgesia after low-frequency electroacupuncture as adjunctive treatment in inguinal hernia surgery with abdominal wall mesh reconstruction. *Acupunct Med.*2015;33(5):360-367.

[7] Yilmaz E, Karakaya E, Baydur H, Tekin I. Effect of transcutaneous electrical nerve stimulation on postoperative pain and patient satisfaction. *Pain Manag Nurs*. 2018;S1524-9042(17):30676-30678.

[8] Chiu JH, Chen WS, Chen CH, Jiang JK, Tang GJ, Lui WY, et al. Effect of transcutaneous electrical nerve stimulation for pain relief

on patients undergoing hemorrhoidectomy: prospective, randomized, controlled trial. *Dis Colon Rectum*. 1999;42(2):180-185.

[9] Yeh ML, Chung YC, Hsu LC, Hung SH. Effect of transcutaneous acupoint electrical stimulation on post-hemorrhoidectomy-associated pain, anxiety, and heart rate variability: a randomized-controlled study. *Clin Nur Res*. 2018;27 (4):450-466.

[10] Mi Z, Gao J, Chen X, Ge Y, Lu K. Effects of transcutaneous electrical acupoint stimulation on quality of recovery during early period after laparoscopic cholecystectomy. *Zhongguo Zhen Jiu*. 2018;38(3):256-260.

[11] Conn IG, Marshall AH, Yadav S, Daly JC, Jaffer M. Transcutaneous electrical nerve stimulation following appendicectomy: the placebo effect. *Ann R Coll Surg Engl*. 1986;68(4):191-192.

[12] Lee S, Nam D, Kwon M, Park WS, Park SJ. Electroacupuncture to alliviate postoperative pain after a laparoscopic appendectomy: study protocol for a three-arm, randomised, controlled trial. *BMJ Open* [Internet] 2017;7. Available at : https://www.ncbi.nlm.nih.gov/pmc/articles/PMC5724078/pdf/bmjopen-2016-015286.pdf

[13] Kim G. Electroacupuncture for postoperative pain and gastrointestinal motility after laparoscopic appendectomy (AcuLap): study protocol for a randomized controlled trial. *Trials* [Internet] 2015;16. Available at: https://www.ncbi.nlm.nih.gov/pmc/articles/PMC4606555/pdf/13063_2015_Article_981.pdf

[14] Bjersa K, Jildenstaal P, Jakobsson J, Egardt M, Faqevik-Olsen M. Adjunct high frequency transcutaneous electric stimulation (TENS) for postoperative pain management during weaning from epidural analgesia following colon surgery: results from a controlled pilot study. *Pain Manag Nur*. 2015;16(6):944-950.

[15] Kim KH, Kim DH, Bae JM, Son GM, Kim KH, Hong SP, et al. Acupuncture and PC6 stimulation for the prevention of postoperative nausea and vomiting in patients undergoing elective

laproscpic resection of colorectal cancer: a study protocol for a three-arm randomised pilot trial. *BMJ Open* [Internet] 2017;7. Available at: https://www.ncbi.nlm.nih.gov/pmc/articles/PMC522 3651/pdf/bmjopen-2016-013457.pdf.

[16] Chen Y, Yang Y, Yao Y, Dai D, Qian B, Liu P. Does transcutaneous electrical acupoint stimulation improve the quality of recovery after thyroidectomy? A prospective randomized controlled trial. *Int J clin Exp Med*. 2015;8(8):13622-13627.

[17] Park C, Choi JB, Lee YS, Chang HS, Shin CS, Kim S, et al. The effect of intra-operative transcutaneous electrical nerve stimulation on posterior neck pain following thyroidectomy. *Anaesthesia.* 2015;70(4):434-439.

[18] Tokuda M, Tabira K, Masuda T, Nishiwada T, Shomoto K. Effect of modulated-frequency and modulated-intensity transcutaneous electrical nerve stimulation after abdominal surgery: a randomized controlled trial. *Clin J Pain*.2014;30(7):565-570.

[19] Sun K, Xing T, Zhang F, Liu Y, Li W, Zhou Z, et al. Perioperative transcutaneous electrical acupoint stimulation for postoperative pain relief following laparoscopic surgery: a randomized clinical trial. *Clin J Pain.* 2017;33(4):340-347.

[20] Xie YH, Chai XQ, Wang YL, Gao YC, Ma J. Effect of electro-acupuncture stimulation of Ximen (PC4) and Neiguan (PC6) on remifentanil-induced breakthrough pain following thoracal esophagectomy. *J Huazhong Univ Sci Technolog Med Sci.* 2014;34(4):569-574.

[21] Bjersa K, Andersson T. High frequency TENS as a complement for pain relief in postoperative transition from epidural to general analgesia after pancreatic resection. *Complement Ther Clin Pract.* 2014;20(1):5-10.

[22] Chatrath V, Khetarpal R, Kumari H, Kaur H, Sharma A. Intermittent transcutaneous electrical nerve stimulation versus transversus abdominis plane block for postoperative analgesia after infraumbilical surgeries. *Anesth Essays Res.* 2018;12(2):349-354.

In: Uses of Electrical Stimulation...
Editor: Jaime Ruiz-Tovar
ISBN: 978-1-53615-036-0
© 2019 Nova Science Publishers, Inc.

Chapter 7

ELECTRICAL STIMULATION FOR IMPROVING PHYSICAL FITNESS PRE- AND POSTOPERATIVELY AFTER ABDOMINAL SURGERY

Artur Marc Hernández[*]
Laboratory of Training Analysis and Optimization,
Sports Research Center, Miguel Hernández University,
Elche, Alicante, Spain

ABSTRACT

In the last few decades, surgical techniques have been developed in terms of safety and efficacy, however, abdominal surgery continues to present some disadvantages. On one hand, is not unusual that patients develope perioperative complications, wich can lead to longer hospital stay, higher costs to health system or even the death of the patient. On the other hand, it achieves bed rest after surgery, which causes a decrease in

---
[*] Corresponding Author Email: arturmarc.hernandez@gmail.com.

muscle mass and muscle strength, and reduces cardiorespiratory fitness. As a consequence, the functional capacity of the patient decreases, so the patient's recovery is delayed.

At present there are numerous studies showing that physical fitness at the time of surgery is a significant factor that relates to postoperative complications, indicating that poor physical fitness is associated with these complications and slower recovery after surgery. Consecuently, in order to prevent these disadvantages, patients need to improve their physical fitness prior to surgery. Although exercise is an effective method for achieving these improvements, in clinical setting, electrical stimulation has several advantages. In sedentary subjects or patients with a poor or impaired physical fitness, it allows strength levels and muscular mass to be increased in a short period of time. In addition, due to its application, d no voluntary physical movement of the patient is required, it can be applied in bed rest conditions. In this case, it has been proven to maintain levels of strength, muscular mass and funtional capacity, resulting in earlier postoperative recovery.

Keywords: NMES, postoperative complications, prehabilitation, muscle strength

INTRODUCTION

Abdominal Surgery

Actually, the number of surgeries is increasing all over the world, especially in developed countries. It has been estimated that between 2004 and 2012, the number of surgeries performed has raised in a 33.6%, and becoming 312.9 million of surgeries performed in 2012 [1]. Among all these surgical procedures, abdominal surgery is one of the most frequently performed surgery, so that it represents 32.5% of the total surgeries [2]. Abdominal surgery refers to any procedure that implies the surgical management of the abdomen or any of the abdominal organs [3]. Abdominal surgery includes colorectal surgery, hepatic and renal transplant, gastrectomy, esophagectomy or bariatric surgery, among other procedures.

Traditionally, this procedure required opening the abdomen through a large incision, however, at present, technological advances have allowed new techniques to emerge, such as laparoscopy, which are less invasive. Compared to open surgery, the use of this approach has shown, among other benefits, a lower mortality rate [4–6], a decreasing number of perioperative complications [5, 7, 8] and an earlier postoperative recovery [9].

Despite the benefits obtained by laparoscopy, abdominal surgery might still present complications. It has been reported that nearly 35.6% of patients suffer from any type of complication after major abdominal surgery, with 20.8% of major complications, which require invasive treatment or admission to intensive care [10]. Another study that was conducted with approximately 466,000 patients, strengthen the latter percentage, concluding that the risk of suffering from one major postoperative complication after intra-abdominal surgery was 18.6% [11]. These complications may provoke great disadvantages. On one hand, it may raise the mortality rate after surgery [12], which in abdominal surgery ranges between 2.73 and 3.5% [2, 10]. For instance, Sabaté et al. established in their study that, among patients who presented postoperative pulmonary complications, mortality rate was around 24.4%, while patients who did not present this type of complication, mortality rate was only 1.2% [13]. On the other hand, these complications may lead to a higher hospital stay [12, 14–16] and the admission to intensive care [12]. Altogether, this lead to an increase in healthcare costs. Some studies have reported that, after major abdominal surgery, myocardial infarction increases the costs in approximately $20,000 per patient [14], while pulmonary complications raise it in around $28,000 [17].

Due to these reasons, it is necessary to detect, determine and understand factors that influence the perioperative complications, in order to decrease the risks of suffering them.

Abdominal Surgery Complications

As previously mentioned, it is not unusual that certain complications may occur after abdominal surgery. Among all of them, the most frequent are infections with 13.3% of cases [18], pulmonary complications with 5.8% of the total patients [19] and cardiac complications, which cause 1.4% of the mortality after major abdominal surgery [14].

There are certain factors that are going to influence on the possibility of developing postoperative complications. Some of them can not be modified, such as age [4, 16, 20], male gender [4, 21] or history of cancer [20]. Nevertheless, there are some other factors, such as the presence of different comorbidities (hypertension, cardio-pathology or diabetes mellitus) [16, 22, 23], smoking [16, 20, 24], body mass index [20, 25], sarcopenia [26–28] and poor physical fitness [21, 22, 24, 29], which are modifiable factors.

Currently, factors related to physical fitness are becoming important on surgery, since they are highly associated with postoperative complications. In fact, some studies determine that these factors are the most strongly related to postoperative complications [21]. Therefore, the improvement of physical fitness may lead to a reduction of these complications and an earlier recovery after surgery.

PHYSICAL FITNESS

Physical fitness has been defined as "a state characterized by (a) an ability to perform daily activities with vigor, and (b) demonstration of traits and capacities that are associated with low risk of premature development of the hypokinetic diseases (i.e., those associated with physical inactivity)" [30]. In general, physical fitness is defined by different components. On one hand, the skill-related components refer to athletic capacities, as agility, power or coordination [31]. On the other hand, the health-related components, are integrated by cardiorespiratory and muscular endurance, muscle strength, body composition and flexibility [31]. Health-related components have a greater incidence on public health, since they are related to the health of individuals. For instance, cardiorespiratory fitness is inversely proportional to mortality [32], while a low muscular strength is directly proportional to premature death [33].

At this point, it is necessary to define several concepts referred to health-related components which are proved to influence on the results of abdominal surgery. In respect of body composition, it is highly remarkable its significant impact on health, since a high weight, especially a high body mass index, is related to different cardio metabolic pathologies [34], and the reduction of life expectancy. In fact, body mass index is associated with mortality risk [35]. Nonetheless, apart from the total weight, it is also important the body composition. Our organism is composed of different tissues (Figure 1), which have an influence on individuals' health status, emphasizing two of them: fat mass and skeletal muscle. The former affects to health status, because an excess of it is related to health risks [36]. The latter, skeletal muscle, presents great benefits for health, since it is associated to an optimal functional capacity and is considered as a secretory organ. In this sense, skeletal muscle has different functions, such as absorbing 75% of the postprandial glucose [37] or releasing of myokines [38].

Here sarcopenia becomes particularly important. Sarcopenia is a syndrome that is characterized by a loss of skeletal muscle in addition to a decrease of muscle strength or performance that develops into adverse results as poor quality of life or physical disability [40]. Although, both low muscle mass and low strength levels are independently associated with postoperative complications, when both conditions appear together, as sarcopenia's case, the risk of suffering from any complication increases [26].

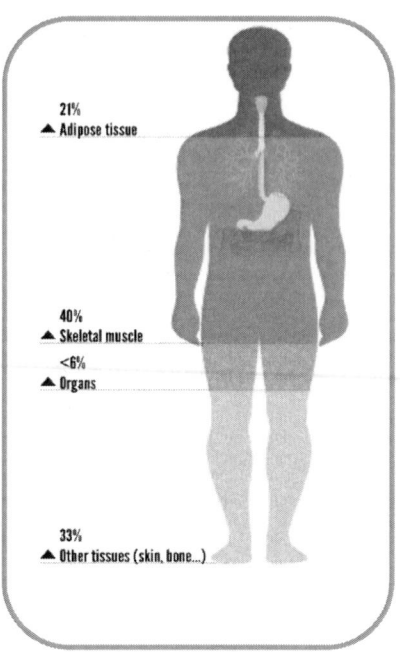

Data obtained from Dullo et al. [39].

Figure 1. Distribution of body tissues.

On the other hand, cardiorespiratory fitness makes reference to the capacity of circulatory and respiratory systems in order to provide fuel during physical activity [31]. Determining the cardiorespiratory fitness in patients undergoing abdominal surgery is really important, since it indicates the cardiorespiratory reserve that every patients disposes, and thus, it can be recognized the risk of suffering from any complication [41]. Cardiopulmonary exercise testing is a suitable method for determining this capacity [42]. This test allows numerous variables to be obtained, however, maximum oxygen uptake (VO_{2max}), peak oxygen uptake (VO_{2peak}) and anaerobic threshold are the most frequent variables for establishing relationships between cardiorespiratory fitness and postoperative complications. The VO_{2max} and VO_{2peak} define the maximum amount of oxygen that the patient is able to use, however it must be noticed that for being considered as VO_{2max}, it must have been produced a plateau in oxygen uptake [43]. In turn, when the oxygen provided to the muscles is not enough to cover the energy demands, anaerobic glycolysis intervenes transcendentally to provide ATP, accumulating lactate exponentially, process known as anaerobic threshold [44].

Although cardiopulmonary exercise testing is the "gold standard" to determine cardiorespiratory fitness, sometimes, the necessary tools to perform this test are not available. Nevertheless, there are other type of measurements, as six-minute walk test, that can be used, since they establish functional capacity of the patients [43] and are also related to postoperative complications [45].

Role of Physical Fitness in Abdominal Surgery Complications

At present, physical fitness is considered itself as a factor of great importance in patients awaiting for abdominal surgery, since it may contribute to reduce perioperative complications.

First of all, there are factors related to muscular function. Sarcopenia has been proved to be really relevant, since it increases between 3.12 and 4.5 times the risk of suffering from postoperative complication [26, 27], so it is shown as an independent predictor of infection (odds ratio: 4.6) [46]. In fact, it has been reported that after colorectal cancer surgery, 23.1% of the patients with sarcopenia suffered from infections, while this data was reduced to 12.6% in patients without sarcopenia [46]. Furthermore, both the incidence of medical and total complications increase according to the level of sarcopenia (Figure 2) [47]. Nevertheless, components, that define sarcopenia (muscle mass and strength), are also related to postoperative complications. Patients with low muscle mass have 3.8 times more chances of suffering from postoperative complications, while patients with low strength levels have proved to have 2.1-2.5 more times of suffering from complications [21, 47].

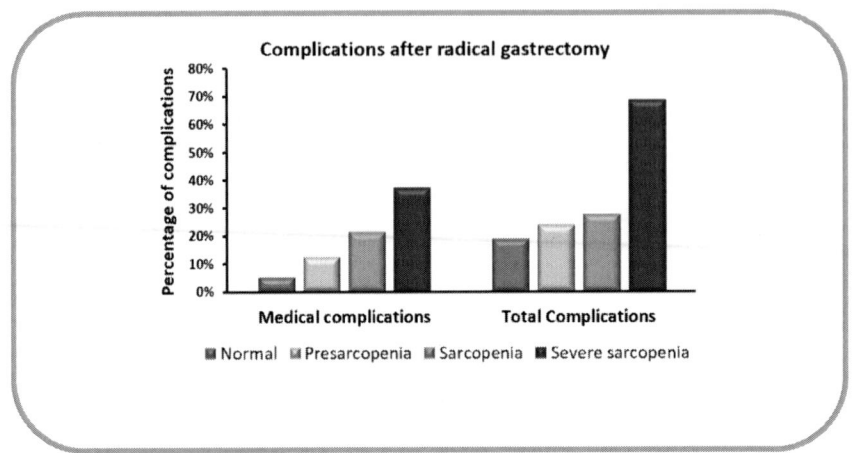

Data obtained from Huang et al. [47].

Figure 2. Percentage of medical and total complications according to sarcopenia level.

On the other hand, cardiorespiratory fitness has proved to be especially significant in postoperative complications. Recent systematic reviews have shown that cardiorespiratory fitness predicts postoperative complications after different major abdominal surgeries, like hepatic transplant, abdominal aortic aneurysm repair, bariatric, upper gastrointestinal and colorectal surgery [12, 48, 49]. Furthermore studies have showed how anaerobic threshold and VO_{2peak} are powerful predictors for different disadvantages, as infections and pulmonary and cardiovascular complications [50]. For example, a VO_{2peak} lower than 15.8 for bariatric surgery and 16,7 ml·kg^{-1}·min^{-1} for major colonic surgery [51, 52] is related to a higher number of complications and a longer hospital stay after bariatric surgery and major colonic surgery [51, 52]. Likewise, a VO_{2peak} < 15 ml·kg^{-1}·min^{-1} predicts mortality after abdominal aortic aneurysm repair [53]. Equally, anaerobic threshold range lower than 9,2-11,4 ml·kg^{-1}·min^{-1} predicts postoperative mortality and morbidity, as well as hospital stay and admission to intensive care unit [54–57]. In addition, poor cardiorespiratory fitness affects to different complications. West et al. showed in their study that 16% and 23% of patients with an anaerobic threshold < 10.1 ml·kg^{-1}·min^{-1} presented pneumonia and infections after surgery, while in patients with a higher anaerobic threshold, these percentages decreased to 3% and 8%, respectively [52].

Besides, the risk of suffering from complications varies depending on different aspects. On one hand, a lower of cardiorespiratory fitness is associated with postoperative complications. It has been found that after major surgery, patients with a VO$_{2peak}$ > 20 ml·kg^{-1}·min^{-1} have a low risk of suffering from complications, while patients with a VO$_{2peak}$ < 15 ml·kg^{-1}·min^{-1} and VO$_{2peak}$ < 10 ml·kg^{-1}·min^{-1} have a moderate and high risk, respectively [58]. On the other hand, these postoperative complications can aggravate if they are accompanied by comorbidities. A study conducted in "elderly patients" shows that, after major abdominal surgery, patients who presented an anaerobic threshold < 11 ml·kg^{-1}·min^{-1} showed a mortality of 18%, but patients who also presented preoperative ischemia, the mortality increased to 42%. In contrast, patients who presented the same morbidity, but with an anaerobic threshold > 11 ml·kg^{-1}·min^{-1}, the mortality rate was 0.8% and 4%, respectively [59].

After the data provided, it can be concluded that a poor physical fitness has a close relationship with postoperative complications. Both body composition and muscular and cardiorespiratory fitness are associated with postoperative complications, finding fewer complications in patients with greater fitness.

SURGICAL STRESS

In order to understand these associations between physical fitness and postoperative complications, it is necessary to know that surgery leads the human body to a state of stress that compromises the organism (surgical stress) [60].

The human body responds to the surgical process through different physiological pathways to maintain homeostasis,, allowing cardiovascular, respiratory and metabolic changes. These changes are, among others, increases in cardiac output, heart rate, blood pressure, oxygen demand, and respiratory rate, as well as a mobilization of energy reserves to provide substrates as fuel to the repair processes of tissue and protein synthesis for the immune response [61, 62]. This will increase the risk of complications [61], therefore, it is necessary that the organism is prepared to deal with these physiological variations.

Surgical Stress and Physical Fitness

Physical fitness will determine the impact of surgical stress on the patient in different ways. After major surgery, there is an increase of 50% in oxygen consumption at rest, (from 110 ml·min^{-1}·m^{-2} to 170 ml·min^{-1}·m^{-2}), which can be maintained for several days [63], and it is not strange that some patients reach an oxygen consumption of 240 ml·min^{-1}·m^{-2} [59]. Consequently, these increases in oxygen demand will generate a deficit of it [45]. If the human body is not able to provide the oxygen necessary to respond to this O_2 deficit, different postoperative complications can be developed [45]. For example, one study showed that patients who died after surgery had an oxygen deficit averaged at its maximum of 33.2_L·m^{-2}, while this deficit was 21.6 L·m^{-2} in those who only had complications (without mortality) and 9.2 L·m^{-2} in patients who did not suffer any complications [64].

Therefore, it is likely that patients with an adequate cardiorespiratory reserve are able to resist this additional oxygen demand without exceeding their physiological parameters [49], thus reducing the probabilities of suffering from complications. This can be of great importance, especially in surgeries, such as liver transplants, in which there is a high mortality rate, around 10% [2]. In this regard, the cardiorespiratory capacity is so important, that only the increase in 1_ml·kg^{-1}·min^{-1} of oxygen consumption at anaerobic threshold decreases the odds ratio of postoperative complications by 20%, while the increase of 2_ml·kg^{-1}·min^{-1} decreases it by 40% [52].

Likewise, skeletal muscle and muscle strength also seem to be important factors toresist successfully surgical stress. In patients with sarcopenia, a reduction in the expression of myocins is observed, which may decrease the activity of NK cells [65], increasing the risk of infection [21]. In fact, both sarcopenia and low levels of fat free mass are associated with the risk of infection after major surgery [46, 66]. For instance, after cardiac surgery, 18.5% of subjects with a low fat free mass index suffer from infectious complications, compared to 4.7% of subjects with a normal fat free mass index [66]. A reason could be that the amount of muscle mass is associated with the reserves of proteins available to the human body. Patients with low protein reserves have been associated with an increased risk of complications (especially pneumonia) and mortality after major abdominal surgery [67]. Thus, a larger amount of skeletal muscle indicates a greater reserve of muscle proteins, which are necessary to respond appropriately to surgical stress [68].

In addition, skeletal muscle could also play a relevant role in different metabolic aspects. Due to surgical stress, after surgery there is a large increase in blood glucose levels [69], which can be maintained for a few days. This can represent a drawback, since impaired fasting glucose is associated with perioperative cardiovascular events in patients undergoing major surgery [70]. Skeletal muscle is considered the largest store of glucose in the body, storing approximately 80% of glucose (around 500 grams) [71]. Therefore, it can be hypothesized that increasing muscle mass before surgery may provide a larger body's ability to store glycogen. This could prevent or reduce postoperative hyperglycemia, thus decreasing the problems associated with it.

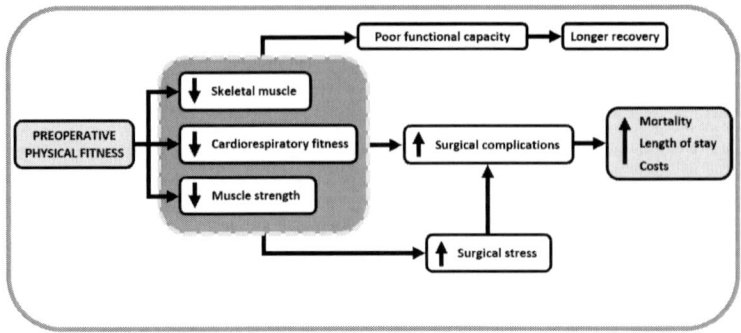

Figure 3. Effects of preoperative fitness on surgical complications.

Relationship between Surgical Stress, Fitness and Recovery

Physical fitness plays a fundamental role in patient's recovery. After a major surgery, human body takes some time to recover functional capacity prior to surgery. After 3 weeks of colon resection surgery, only 41% of the patients showed equal or higher values of functional capacity compared to the preoperative levels [72]. However, patients who increase their functional capacity prior to surgery show an earlier recovery of walking capacity (Figure 4) [72]. For example, at two months of colorectal surgery, 81% of the patients who performed a prehabilitation program that included a part of exercise, recovered their functional capacity levels prior to surgery, compared to 40% of patients who did not perform this program [73].

Both low levels of strength and muscle mass have been associated with a longer hospital stay after major surgery [66], therefore, it is necessary to reach surgery with these variables in an optimal state. Due to surgical stress, protein degradation is accelerated, in turn protein synthesis is low, which leads to a negative protein balance, causing the degradation of muscle tissue [62]. Moreover, abdominal surgery is followed by a period of inactivity, which causes losses of muscle mass and muscle function, and as a consequence, recovery after surgery would be delayed.

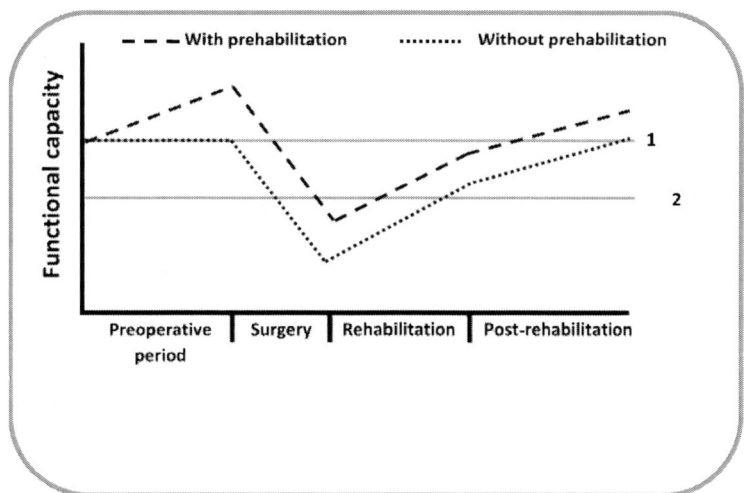

Adapted from Santa Mina et al. [77]. 1, initial functional capacity; 2, functional threshold.

Figure 4. Effects of increase functional capacity prior to surgery on the functional recovery after surgery, theorical model.

A study showed how, after only 10 days of bed rest, there was a reduction of 1.5 kg of lean mass, reducing by 6.3% lean body mass and by 15.6% muscle strength of lower extremity [74]. These values can be aggravated with admission to intensive care, since it has been reported that after 10 days, there have been reductions in rectus femoris cross-sectional area of 17.7%, observing that this loss increases proportionally with days of admission in intensive care [75]. Unfortunately, bed rest also causes decreases in cardiorespiratory fitness. In fact, it has been reported decreases of 7.6% and 15.3% of VO_{2peak} in young andolder people, in just 2 weeks of bed rest [76].[

ELECTRICAL STIMULATION AND PHYSICAL FITNESS

As has been observed, physical fitness is of great importance in abdominal surgeries. Improving physical fitness prior to surgery has two well-defined objectives. On one hand, to provide the organism with an adequate physiological reserve so it is able to cope with the stress induced by surgery. On the other hand, the improvement of physical fitness favors an earlier recovery. In turn, after surgery, a period of bed rest will occur, which will lead to muscle wasting and decrease the functional capacity of the patients.

Electrical stimulation has been shown as an optimal method to achieve improvements in physical fitness, especially in the clinical setting, where it is common to find sedentary or with a poor or impaired physical fitness patients. In addition, it is an effective treatment in bed rest conditions and in intensive care admissions, when voluntary exercise is not able to be applied.

Electrical stimulation is a non-invasive treatment that consists on the delivery of stimuli using electrodes that are placed on the skin [78]. There are different types of electrical stimulation, which vary according to their objectives: a) transcutaneous electrical nerve stimulation, which is used for treating acute and chronic pain conditions [79]; b) functional electrical stimulation, which aim is to help patients to perform a certain function, such as getting out of a chair [80]; c) neuromuscular electrical stimulation (NMES), which is used mainly during rehabilitation periods [81], being a useful tool to improve and maintain physical condition. Therefore, this chapter will focus on this last modality of electrical stimulation, due to the benefits that presents on physical fitness.

Effects of Neuromuscular Electrical Stimulation on Physical Fitness

Body Composition

The main advantage of NMES over body composition lies in its effects on skeletal muscle. To achieve increases in muscle mass by means of conventional resistance training, it is necessary to train with high loads around 60 - 80% of the maximum repetition [82]. However, many of the patients who are undergoing major surgery are sedentary subjects, which means that their body is not prepared to train at high intensities. To reduce the risk of injury, a period of anatomical adaptation before hypertrophy training should be carried out. However, NMES generates increases in muscle mass, especially in sedentary individuals [83], with the advantages that brings as a safe treatment, and therefore does not require such a long adaptation period. These characteristics make this method an effective alternative to use prior to surgery.

Regarding the intervention time that is necessary to achieve hypertrophy by NMES, although some studies have reported that it is possible to obtain hypertrophy in 4 weeks [84], most show that it takes around 8 weeks. Gondin et al. reported that after 4 weeks of NMES there were no significant increases in quadriceps cross-sectional area, finding significant gains after 8 weeks of training [85]. Likewise, other studies have shown increases of approximately 10% cross-sectional areal of the quadriceps femoris after 9 weeks of electrical stimulation in sedentary subjects [83].

One of the most remarkable advantages of NMES are its safety and effectiveness in patients with weakened muscles or under special situations, such as bed rest after surgery or admission in intensive care, where patients' physical fitness is reduced. At this point, NMES assumes particular relevance, since it avoids reductions in muscle mass [86]. For instance, it has been reported that after a period of 6-7 weeks of immobilization of the leg, the group that received treatment with NMES had a reduction of only 4.5% of quadriceps cross-sectional area, compared to a reduction of 17% of the group without treatment [87].

Unfortunately, some surgical complications result in a patient's admission to intensive care. In this situation, patients may have a reduction of up to 21.5% of sectional diameter of the vastus intermedius, however, in those who receive treatment through NMES, this loss is reduced by almost half, being 12.5% [88]. In addition, this is achieved safely, since the intervention by NMES in patients in intensive care does not cause increases in heart rate, blood pressure or O_2 saturation [89], indicating that it does not suppose a stressful treatment for the organism. Apart from that, NMES is considered a useful method to recover muscle mass during rehabilitation periods. In patients who have decreased their muscle mass, NMES has generated raises in the total area of the quadriceps (from 129.2 cm^2 to 142.2 cm^2) after 12 weeks of intervention [90].

Although treating aspects focused on nutrition is not the objective of this chapter, it must be emphasized that in order to achieve the desired effects on muscle architecture (gain and maintenance of muscle mass), it is essential to have an adequate protein intake. To achieve increases in muscle mass it will be necessary to intake about 1.8 grams of protein per kilogram of weight, while the intake of 1.4 and 1.8 grams per kilograms of weight can prevent muscle protein catabolism, and therefore avoid the loss of muscle mass [91]. In fact, some studies recommend to take 25-30 grams of high-quality protein at each meal in order to minimize the loss of muscle mass during bed rest [92].

Muscle Strength and Functional Capacity

At present, it is widely accepted that NMES improves strength levels. The studies show increases of between 7.7% and 60% of the maximum dynamic force [83, 93], although it should be noted that the treatment must be carried out dynamically. Regarding isometric strength, these increases can reach up to 47.5% for the brachial biceps [94], 44% for the quadriceps [95], or 58% in abdominal strength [96]. Equally, NMES is effective for improving muscular endurance, showing increases of up to 243% after treatment [97]. Furthermore, these increases can be obtained in a short period of time. Although most studies last for 8 weeks or longer, shorter studies or intermediate evaluations also show positive effects on strength levels. For example, in only 2 weeks of treatment with NMES, there is 64% increase in muscular endurance [97], while after 5 weeks of treatment 33% increase in maximal isometric strength is achieved [93].

These benefits can be transferred to the clinical setting. Patients who are in a period of immobilization show a muscle strength loss of approximately 1.8% per day [86]. However, it has been widely accepted that NMES avoids or reduces the loss of strength in these situations, and also contributes to the recovery of strength levels during the rehabilitation period [98, 99]. Similarly, treatment with NMES avoids strength reductions in patients in intensive care [100], enabling a recovery of force 4.5 times faster compared to subjects who are not on treatment [101].

The loss of strength that will be generated after surgery will cause reductions in functional capacity [86], which is of vital importance to perform day-to-day activities. But NMES has also proved to be an adequate method to prevent these reductions. In adults with advanced disease, through 6 weeks of NMES the results in the six-minute walk test (+ 35 m), incremental shuttle walk test (+ 9 m) and endurance shuttle walk test (+ 64 m) are improved [102]. Likewise, patients in the rehabilitation period, where NMES was applied for 12 weeks, achieved a reduction in the time to perform the "sit to stand to sit" test and an improvement of approximately 10% of the six-minute walk test [103].

Cardiorespiratory Fitness

Many of the studies in which electrical stimulation is used for improving cardiorespiratory fitness are performed in combination with exercise, obtaining enhancements in VO_{2peak} [104, 105]. However, when electrical stimulation is applied in isolation, this technique does not seem to be so effective. Currently, this topic has been addressed by several meta-analyses. In one of them, it was determined that subjects who received NMES, achieved improvements in VO_{2peak}, but these were trivial (+ 0.10 $L \cdot min^{-1}$) [106], while another meta-analysis showed that NMES did not produce increases in VO_{2peak} [102]. However, two important factors must be considered On one hand, it is possible that parameters of NMES have not been correctly handled, and because of it, improvements have not been achieved. On the other hand, it should be noted that these analyzes were performed in patients who were clinically stable [106]. Possibly, NMES is effective in patients with poor or impaired cardiorespiratory fitness. As has been reported, after the application of NMES, healthy subjects only improved their functional endurance by 4%, however, clinical populations had improvements of 35% [107].

NMES presents different advantages in the clinical setting. First, it can be used for patients who are unable to exercise voluntarily. Apart from that, the movement caused by NMES is minimal, [108], in that way, it is unlikely to affect tissues that have been damaged during surgical intervention. Therefore, the use of NMES may be relevant to maintaining or increasing VO_{2peak} under clinical conditions, like bed rest or rehabilitation phase. For example, after cardiac transplant, through 8 weeks of NMES, VO_{2peak} was increased from 17.1 to 18.7_ml·kg^{-1}·min^{-1} [109].

Limitations of Neuromuscular Electrical Stimulation

Nevertheless, treatment with NMES to improve physical fitness has some limitations, especially when compared to conventional exercise. First, the order of fiber recruitment. When a voluntary contraction occurs, small motor units (composed of slow fibers) are recruited in first place, and progressively larger motor units (composed of fast fibers) are recruited too [110]. This theory is known as "size principle" [110]. This order differs when electrical stimulation is used [98], presenting a disordered recruitment, stimulating both large and small motor units from the beginning [111]. This behaviour presents a great disadvantage. As large motor units are composed of fast fibers, that are more fatigable [112], NMES presents an earliest emergence of the fatigue compared to voluntary contraction [111].

Unfortunately, NMES has more drawbacks. This method is not able to activate all muscle fibers, activating those that are closest to the electrodes, that is to say, the superficial fibers, and not reaching the deepest ones [113]. In addition, by NMES, the same motor units are always recruited, which will contribute to the emergence of early fatigue and therefore a decrease in torque [114]. Finally, when a voluntary contraction of a certain duration occurs, other muscle fibers are activated to replace those which are fatigued. This effect does not occur in NMES, in which the same muscle fibers are activated throughout the contraction (simultaneous recruitment), causing a greater fatigue [115].

All these factors will provoke that NMES causes great fatigue. On the other hand, for NMES to be effective, it is necessary to work at high intensities, reaching the threshold of pain, which will generate discomfort, reducing tolerance and adherence to treatment [116].

Maximal Voluntary Contraction

A fundamental concept in the NMES treatment is the maximal voluntary contraction (**MVC**), which is the maximum force that an individual is able to perform in a specific isometric exercise [117]. MVC is useful both to evaluate strength improvements after treatment with NMES and determine the intensity training. Therefore, NMES intensity training is the ratio between MVC force and NMES evoked force, expressed in percentage [118]. For example, if a subject has a maximum force of 200 Newton, and the treatment with NMES reaches 100 Newton, this means that subject is training at 50% of the MVC. An important aspect is that muscle activation increases linearly to the MVC percentage. In fact, even a formula has been proposed to determine quadriceps muscle cross-sectional area activated by NMES:

$$\text{Activated area (\%)} = 0.70 \cdot \% \text{ MVC} + 0.77 \text{ [119]}.$$

The effectiveness of the NMES treatment depends on the MVC percentage at which training is carried out. It has been observed that if high muscle tension is achieved [120], the effects on hypertrophy improve. Equally, strength gains are associated with the intensity of treatment [121]. For instance, in one study it was reported that subjects who performed the treatment of NMES at 25% of the MVC, increased their strength levels by 24%, while subjects who reached 50% in the treatment had gains of 48% [122].

Parameters of Neuromuscular Electrical Stimulation

To optimize the treatment through NMES, several aspects must be considered. On one hand, high percentages of the MVC must be reached, trying to stimulate as many muscle fibers as possible. On the other hand, the appearance of fatigue must be minimized and the sensation of discomfort must be reduced. This will increase the torque during the sessions and let a greater tolerance and adherence to the treatment. In order to maximize the benefits of NMES, the different parameters that constitute this method must be manipulated correctly. In the following paragraphs, the most important parameters will be described (Figure 5).

Frequency

The frequency is expressed in Hertz (Hz), and refers to the "pulses" that are repeated in a second. As previously seen, to be effective, the treatment with NMES must cause a high percentage of the MVC, and this will increase with the frequency used. Unfortunately, the frequency is directly associated with the fatigue caused, observing that the higher the frequency used, the greater the fatigue obtained [123]. In fact, the frequency has been more decisive in the appearance of fatigue compared to other parameters, such as current amplitude or pulse duration [124]. This will decrease the torque that is reached during the session, and therefore, the effectiveness of the treatment. It was observed that by applying a frequency of 100 Hz, the maximum torque was reduced by 76% at the end of the session, while by using a frequency of 25 Hz, the reduction was by 39% [124]. However, some authors have reported that working with frequencies lower than 50 Hz, tetanic contractions are not achieved, which is essential to generate benefits in muscle strength and muscle architecture [125]. In turn, too high frequencies will generate great fatigue, reducing the effectiveness of the treatment.

Nevertheless, when using high frequencies and occurring tetanic contractions, cardiovascular response is minimal, according to some authors, is approximately of 2 metabolic equivalents [126]. However, low frequencies produce sub-tetanic isometric muscle contractions, which generate a greater oxygen demand in the tissues [127]. Thereby, oxygen consumption raises and is able to reach into 53% of the VO_{2max} [128] or requires a cardiovascular response between 4 - 7 metabolic equivalents [127].

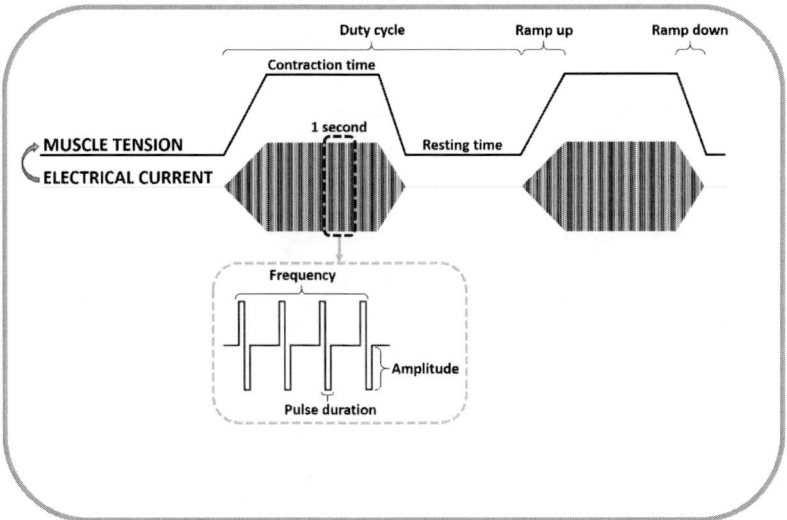

Adapted from Vivodtzev et al. [125].

Figure 5. Electrical current parameters and theorical model of muscle tension evoked by electrical current.

Therefore, the frequency used will depend on the objective of the intervention. In order to maintain or recover skeletal muscle mass and muscle strength, it is recommended to apply frequencies between 50-120

Hz [98, 125, 129, 130]. On the contrary, to improve cardiorespiratory fitness, it is recommended to use low frequencies, between 4-10 Hz [108, 127, 128, 131], where 4 Hz frequencies has been proved to be especially effective [126].

Pulse Duration

Pulse duration, also known as pulse width is the duration of each current pulse, which is usually expressed in microseconds (μs). This parameter will influence directly the results of the treatment by electrical stimulation, since the higher the pulse duration applied, the greater the torque obtained [1]. For example, increasing the pulse duration from 150 to 450 μs generates an increase of 40% of the muscle activation and enlarges torque by 55% [2]. In addition, a longer pulse duration will allow to recruit more fibers and reach deeper tissues, and it will be also useful when it is pursued to treat layers of secondary tissue [3]. However, too high pulse duration (> 500 μs) should not be used, since sensory nerve fibers will be recruited, which causes a large discomfort. On the contrary, too short pulse duration (< 100 μs) will not generate benefits [4]. This parameter manifests a great advantage. As long as it is within the appropriate ranges (between 150 - 500 μs), the increases in pulse duration do not increase fatigue [5] and are well tolerated by subjects [6]. Thus, it allows more efficient interventions.

Lastly, some authors have reported that pulse duration will depend on the size of the muscles to be treated. For larger muscle groups, greater pulse duration (> 300 μs) should be used, whereas lower pulse duration (100 - 300 μs) are recommended for smaller muscle groups [7]. In summary, with the aim of achieving high muscular tension, with a low fatigue index, it has been recommended to use pulse duration between 100 and 500 μs [8, 9].

Duty Cycle

Duty cycle is the relation that exists between the time in which the stimulation is activated, that is, when muscle contraction is provoked, and the period in which it is deactivated. In the literature it is usually expressed in two different ways: a) as a ratio, for example 1:5, which means that muscle contraction is 1 second-long and resting time is 5 second-long; b) in percentage, for example 20% indicates that the stimulation is activated for 20% of a complete cycle. This parameter affects directly fatigue, and therefore, could reduce the effectiveness of the treatment. As the study by McLoda et al. showed, when using a duty cycle of 10%, the average of forced evoked was 132.9 Newton, while using a duty cycle of 90%, this average decreased to 41.3 Newton [132]. Equally, it can be observed how 72% of the subjects who used a duty cycle with a ratio of 1:1, decreased their initial force output by 50% in the first 10 minutes of electrical stimulation, while using a ratio of 1:5, only 6% of sample showed this fatigue [133]. Therefore, to reduce the fatigue caused by NMES within the same session, an effective strategy would be to extend the resting time in the duty cycle [134].

Nevertheless, duty cycle will be different depending on the desired objectives: a) to achieve increases in muscle mass, some authors recommend to use high duty cycles (1:1 or 1:2), with contractions of at least 2 seconds [120]; b) to obtain an increase in force, it must be used lower duty cycles, such as 1:5 [135]; c) concerning cardiorespiratory fitness, it is required high duty cycles, with ratios between 1:1 and 9:2 in order to obtain improvements [131, 136].

Amplitude

The amplitude is the intensity with which current is applied. It is usually expressed in milliamperes (mA), although sometimes it also appears in millivolts (mV). It is one of the most important parameters of electrical stimulation, since the torque produced increases linearly with amplitude [137]. In fact, according to some authors, muscle tension generated by amplitude is the greatest determinant of the effectiveness of NMES [138].

Treatment with high amplitudes has concrete advantages. NMES treatment stimulates the superficial muscle fibers, which are closer to the electrodes, observing how the recruitment of fibers decreases when distanced from the electrodes [113]. However, when using higher amplitudes, apart from superficial fibers, deeper muscle fibers are depolarized, which are further from the electrodes [139]. As a result, a greater amplitude provokes a greater activation of the muscle and, therefore, a greater resulting torque [140], reaching higher percentages of the MVC [141]. In addition to obtaining a greater torque, the application of a higher amplitude will let the organism to enlarge the VO_{2max}. For instance, using an ~86 mA amplitude allows to reach 40% of the VO_{2max}, however, when rising amplitude to ~ 104 mA, it is achieved 53% of the VO_{2max} [108]. This intensity training has proved to be effective in improving cardiorespiratory fitness [128].

Furthermore, the amplitude has not been proved as a decisive parameter in the development of fatigue. For example, one study showed that frequency influences more than amplitude in fatigue [142], while it has been reported that increasing the amplitude enlarges the torque produced without increasing fatigue during NMES [124]. The major drawback of using high amplitudes resides in patient tolerance, since high amplitudes generate discomfort [143]. Therefore, the amplitude used depends on the tolerance of the patient [144], where it is recommended to apply the maximum amplitude that is tolerated in order to optimize the treatment [116].

There are different strategies that will help to increase tolerance and allow to train at higher amplitudes. In first place, training should not start using the highest amplitudes that the patient is able to bear, since this produces a certain degree of pain and discomfort, generating reticence of the patient to use electrical stimulation and, therefore, a low adherence [95]. Secondly, it must be considered that, at the beginning of NMES, the maximum amplitude that is tolerated increases rapidly during the first 10 days of treatment, after which a plateau occurs [145]. Therefore, before beginning a treatment with NMES, a period of 8-10 days of familiarization should be carried out. Thirdly, in order to maximize the effects of NMES, it is advisable to increase the intensity within the same session, since superficial muscle fibers (closer to the electrodes) can be fatigued [119]. As shown, increasing the amplitude allows stimulate deeper fibers, thus maintaining MCV levels [119].

Electrodes: Size and Location

Both size and location of the electrodes will determine the effectiveness of an NMES treatment. Firstly, the size of the electrodes is relevant in different aspects. Very small-sized electrodes will have a higher current density, which will increase the sensation of pain, generating discomfort [146]. On the contrary, large-sized electrodes are more comfortable, at the same time produce a greater torque [146]. However, the use of electrodes with an excessively large size should be avoided, since they could stimulate other muscles, which would lead to a loss of efficacy of the treatment [145].

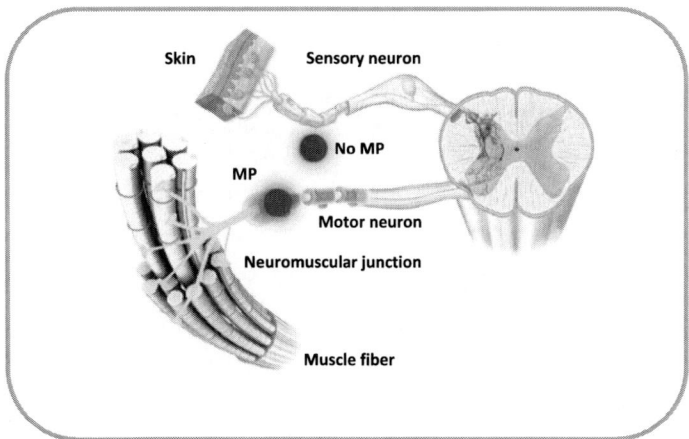

Adapted from Gobbo et al. [148]. MP, motor point.

Figure 6. Location of motor point.

Although there has been some debate about the location of the electrodes, it has already been reported that placing the electrodes on the motor point generates a greater torque, in addition to decreasing discomfort sensation [147]. The motor point is the region that covers the peripheral neuromuscular junctions, so it represents the closest area of the skin to the motor end-plates of the cross-sectional area of the muscle, which are highly excitable [98]. When placing the electrode on motor point, a less intense current is required in order to excite the motor axons and generate muscular contraction [148]. On the contrary, when placing the electrode in a point different from motor point, the intensity to reach the motor branch must be greater, increasing the possibility of exciting sensory fibers, which transmit the pain (Figure 6) [148].

Although currently there are atlas that indicate where the motor points of each muscle are located, the situation of these points is different for each subject [149], so it is necessary to locate them individually, which is quite simple. Only one pen-electrode is needed, with which the surface of the muscle is mapped until one concrete point is found. This point is the one that generates a contraction with the minimum possible current intensity [149]. In turn, the electrodes should be placed longitudinally with respect to the muscle fibers, because this position will generate a greater torque compared to transverse position [150].

Finally, some authors recommend, with the aim of activating other muscle fibers and minimizing fatigue, to change the electrodes of position, both in the same session and between different sessions [118]. Furthermore, as can be seen in Figure 7, a muscle may have several motor points that are going to activate different muscle portions [152]. This fact strengthens the usefulness of changing the position of the electrodes. While other authors have determined that using a greater number of electrodes will allow more motor fibers to be recruited, reducing fatigue [134].

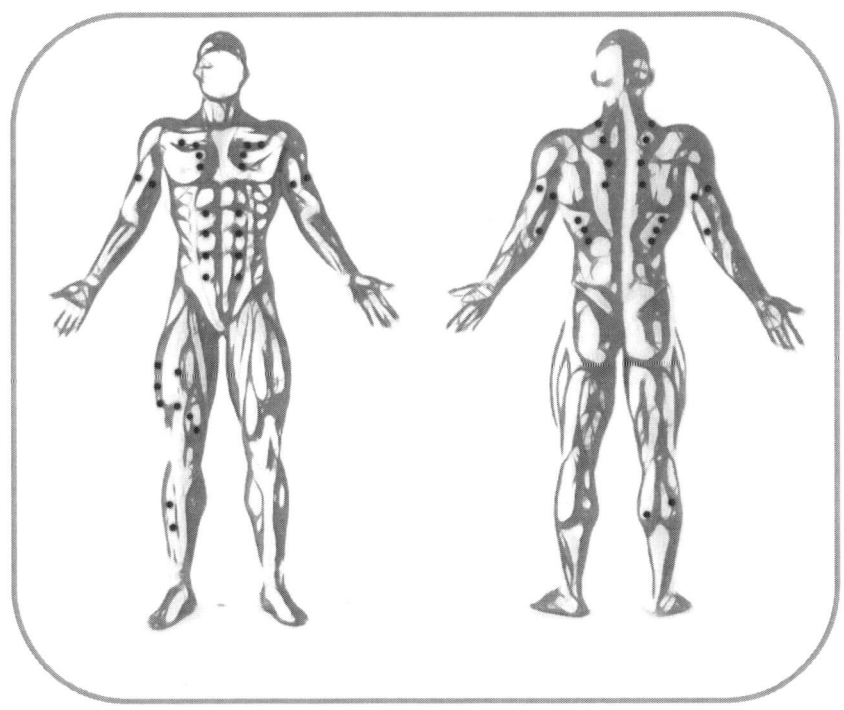

Data obtained from Franz et al. [151] and Botter et al. [149].

Figure 7. Motor points for the upper and lower body.

Treatment Frequency

The frequency of the treatment trough NMES also has an influence on the results. Although it seems that, 2-3 sessions per week are effective in order to achieve strength improvements, several studies have proved how gains increase with volume of treatment. For instance, 3 session training generated greater benefits compared to one of 2 times per week, while 5 session training caused higher gains of strength compared to one of 3 times per week.

In turn, to rise cardiorespiratory fitness, 5 sessions per week (1 hour/session) have proved to be especially effective [81, 127]. Furthermore, enhancements of the VO_{2peak} are significantly associated with training hours [128]. Then, it seems that a higher volume of training will increase the effectiveness of NMES.

Conclusion

Electrical stimulation is a method that offers great advantages in the clinical setting. It is a safe and effective treatment, especially for sedentary subjects. Within the types of electrical stimulation, NMES may be very useful in abdominal surgery, since it includes the following benefits: a) improves the physical fitness of patients prior to surgery, reducing the risk of postoperative complications; b) maintains physical fitness in bed rest conditions; c) allows the patient to increase the level of physical fitness faster, obtaining a patient's early recovery.

However, in order to obtain the highest effectiveness of the treatment, several aspects must be considered. First, the intended aim has to be defined. Second, electrical stimulation is composed of parameters that must be properly manipulated. The most relevant issues to maximize the benefits of the NMES treatment are:

- Training at high percentages of maximal voluntary contraction will generate a greater muscle activation, which provides larger benefits.
- To allow the patient to tolerate high intensities, it is necessary a familiarization period of approximately 10 days.
- In order to improve muscle strength and muscle mass, high frequencies (50 - 120 Hz) must be used, while low frequencies (4 - 10 Hz) just increase the cardiorespiratory fitness.
- Pulse duration must be applied in ranges of values between 100 - 500 μs, which is shorter for small muscle groups (100 - 300 μs) and longer for larger muscle groups (> 300 μs).
- High duty cycles must be used in order to increase muscle mass (1:1 or 1:2) and cardiorespiratory fitness (1:1, 2:9). On the contrary, low duty cycles (1:5) must be applied to achieve strength improvements.

- The maximum voluntary contraction increases linearly with amplitude. Similarly, the increase in this last parameter does not lead to significant increases of fatigue.
- Electrical stimulation effectiveness increases when the electrodes are placed in motor points.
- Fatigue increases with frequency, while modifying pulse duration or amplitude has not a significant effect on the fatigue.
- Regarding the frequency of treatment, 2 - 3 sessions per week develop strength gains and 5 sessions per week is especially effective in order to improve cardiorespiratory fitness.

REFERENCES

[1] Weiser, TG; Haynes, AB; Molina, G; Lipsitz, SR; Esquivel, MM; Uribe-Leitz, T; et al. Estimate of the global volume of surgery in 2012: an assessment supporting improved health outcomes. *Lancet*, 2015, 385, S11.

[2] Noordzij, PG; Poldermans, D; Schouten, O; Bax, JJ; Schreiner, FAG; Boersma, E. Postoperative Mortality in The Netherlands. *Anesthesiology*, 2010, 112, 1105–15.

[3] Mosby. *Mosby's Medical Dictionary*, 10th Edition., 2016.

[4] Masoomi, H; Kang, CY; Chen, A; Mills, S; Dolich, MO; Carmichael, JC; et al. Predictive factors of in-hospital mortality in colon and rectal surgery. *J Am Coll Surg*, 2012, 215, 255–61.

[5] Wang, CL; Qu, G; Xu, HW. The short- and long-term outcomes of laparoscopic versus open surgery for colorectal cancer: a meta-analysis. *Int J Colorectal Dis*, 2014, 29, 309–20.

[6] Rausa, E; Bonavina, L; Asti, E; Gaeta, M; Ricci, C. Rate of Death and Complications in Laparoscopic and Open Roux-en-Y Gastric Bypass. A Meta-analysis and Meta-regression Analysis on 69, 494 Patients. *Obes Surg*, 2016, 26, 1956–63.

[7] Inokuchi, M; Otsuki, S; Ogawa, N; Tanioka, T; Okuno, K; Gokita, K; et al. Postoperative Complications of Laparoscopic Total Gastrectomy versus Open Total Gastrectomy for Gastric Cancer in a Meta-Analysis of High-Quality Case-Controlled Studies. *Gastroenterol Res Pract*, 2016, 2016.

[8] Haverkamp, L; Weijs, TJ; Van Der Sluis, PC; Van Der Tweel, I; Ruurda, JP; Van Hillegersberg, R. Laparoscopic total gastrectomy versus open total gastrectomy for cancer: A systematic review and meta-analysis. *Surg Endosc Other Interv Tech*, 2013, 27, 1509–20.

[9] Aziz, O; Constantinides, V; Tekkis, PP; Athanasiou, T; Purkayastha, S; Paraskeva, P; et al. Laparoscopic versus open surgery for rectal cancer: A meta-analysis. *Ann Surg Oncol*, 2006, 13, 413–24.

[10] Straatman, J; Cuesta, MA; Gisbertz, SS; van der Peet, DL. Value of a step-up diagnosis plan: CRP and CT-scan to diagnose and manage postoperative complications after major abdominal surgery. *Rev Esp Enfermedades Dig*, 2014, 106, 515–21.

[11] Kim, M; Wall, MM; Li, G. Risk Stratification for Major Postoperative Complications in Patients Undergoing Intra-abdominal General Surgery Using Latent Class Analysis. *Anesth Analg*, 2018, 126, 848–57.

[12] Levett, D; Grocott, M. Cardiopulmonary Exercise Testing for Risk Prediction in Major Abdominal Surgery. *Anesth Clin*, 2015, 33, 1–16.

[13] Sabaté, S; Mazo, V; Canet, J. Predicting postoperative pulmonary complications: Implications for outcomes and costs. *Curr Opin Anaesthesiol*, 2014, 27, 201–9.

[14] Sanaiha, Y; Juo, YY; Aguayo, E; Seo, YJ; Dobaria, V; Ziaeian, B; et al. Incidence and trends of cardiac complications in major abdominal surgery. *Surg* (United States), 2018.

[15] Lawrence, VA; Hilsenbeck, SG; Mulrow, CD; Dhanda, R; Sapp, J; Page, CP. Incidence and hospital stay for cardiac and pulmonary

complications after abdominal surgery. *J Gen Intern Med*, 1995, 10, 671–8.

[16] Pereira, EDB; Fernandes, ALG; Anção, M da S; Peres C de, A; Atallah, ÁN; Faresin, SM. Prospective assessment of the risk of postoperative pulmonary complications in patients submitted to upper abdominal surgery. *Sao Paulo Med J*, 1999, 117, 151–60.

[17] Thompson, DA; Makary, MA; Dorman, T; Pronovost, PJ. Clinical and economic outcomes of hospital acquired pneumonia in intra-abdominal surgery patients. *Ann Surg*, 2006, 243, 547–52.

[18] Pessaux, P; Msika, S; Atalla, D; Hay, J; Flamant, Y. Risk Factors for Postoperative Infectious Complications in Noncolorectal Abdominal Surgery. *Arch Surg*, 2003, 138, 314–24.

[19] Yang, CK; Teng, A; Lee, DY; Rose, K. Pulmonary complications after major abdominal surgery: National Surgical Quality Improvement Program analysis. *J Surg Res*, 2015, 198, 441–9.

[20] Brooks-Brunn, JA. Predictors of postoperative pulmonary complications following abdominal surgery. *Chest*, 1997, 111, 564–71.

[21] Sato, T; Aoyama, T; Hayashi, T; Segami, K; Kawabe, T; Fujikawa, H; et al. Impact of preoperative hand grip strength on morbidity following gastric cancer surgery. *Gastric Cancer*, 2016, 19, 1008–15.

[22] Wang, SL; Zhuang, CLe; Huang, DD; Pang, WY; Lou, N; Chen, FF; et al. Sarcopenia Adversely Impacts Postoperative Clinical Outcomes Following Gastrectomy in Patients with Gastric Cancer: A Prospective Study. *Ann Surg Oncol*, 2016, 23, 556–64.

[23] Choudhuri, A; Chandra, S; Aggarwal, G; Uppal, R. Predictors of postoperative pulmonary complications after liver resection: Results from a tertiary care intensive care unit. *Indian J Crit Care Med*, 2014, 18, 358.

[24] Scholes, RL; Browning, L; Sztendur, EM; Denehy, L. Duration of anaesthesia, type of surgery, respiratory co-morbidity, predicted VO2max and smoking predict postoperative pulmonary

complications after upper abdominal surgery: An observational study. *Aust J Physiother*, 2009, 55, 191–8.

[25] Benotti, P; Wood, GC; Winegar, DA; Petrick, AT; Still, CD; Argyropoulos, G; et al. Risk Factors Associated With Mortality After Roux-en-Y Gastric Bypass Surgery. *Ann Surg*, 2014, 259, 123–30.

[26] Huang, DD; Wang, SL; Zhuang, CL; Zheng, BS; Lu, JX; Chen, FF; et al. Sarcopenia, as defined by low muscle mass, strength and physical performance, predicts complications after surgery for colorectal cancer. *Color Dis*, 2015, 17, O256–64.

[27] Peng, PD; Van Vledder, MG; Tsai, S; De Jong, MC; Makary, M; Ng, J; et al. Sarcopenia negatively impacts short-term outcomes in patients undergoing hepatic resection for colorectal liver metastasis. *HPB*, 2011, 13, 439–46.

[28] Jones, K; Gordon-Weeks, A; Coleman, C; Silva, M. Radiologically Determined Sarcopenia Predicts Morbidity and Mortality Following Abdominal Surgery: A Systematic Review and Meta-Analysis. *World J Surg*, 2017, 41, 2266–79.

[29] Reddy, S; Contreras, CM; Singletary, B; Bradford, TM; Waldrop, MG; Mims, AH; et al. Timed Stair Climbing is the Single Strongest Predictor of Perioperative Complications in Patients Undergoing Abdominal Surgery. *J Am Coll Surg*, 2016, 222, 559–66.

[30] Pate, RR. The evolving definition of physical fitness. *Quest*, 1988, 40, 174–9.

[31] Caspersen, CJ; Powell, KE; Christenson, GM. Physical Activity, Exercise, and Physical Fitness: Definitions and Distinctions for Health-Related Research. *Public Heal Rep*, 1985, 100, 126–31.

[32] Lee, DC; Sui, X; Ortega, FB; Kim, YS; Church, TS; Winett, RA; et al. Comparisons of leisure-time physical activity and cardiorespiratory fitness as predictors of all-cause mortality in men and women. *Br J Sports Med*, 2011, 45, 504–10.

[33] Ortega, FB; Silventoinen, K; Tynelius, P; Rasmussen, F. Muscular strength in male adolescents and premature death: Cohort study of one million participants. *BMJ*, 2012, 345, 1–12.

[34] Savva, SC; Lamnisos, D; Kafatos, AG. Predicting cardiometabolic risk: Waist-to-height ratio or BMI. A meta-analysis. *Diabetes, Metab Syndr Obes Targets Ther*, 2013, 6, 403–19.

[35] Kitahara, CM; Flint, AJ; Berrington de Gonzalez, A; Bernstein, L; Brotzman, M; MacInnis, RJ; et al. Association between Class III Obesity (BMI of 40-59 kg/m2) and Mortality: A Pooled Analysis of 20 Prospective Studies. *PLoS Med*, 2014, 11.

[36] Sellen, D. Physical status: the use and interpretation of anthropometry. Report of a WHO Expert Committee. *WHO Tech Rep*, 1995, 854, 1–452.

[37] Houmard, JA; Pories, WJ; Dohm, GL. Severe obesity: evidence for a deranged metabolic program in skeletal muscle? *Exerc Sport Sci Rev*, 2013, 40, 204–10.

[38] Pedersen, BK; Febbraio, MA. Muscles, exercise and obesity: Skeletal muscle as a secretory organ. *Nat Rev Endocrinol*, 2012, 8, 457–65.

[39] Dulloo, AG; Jacquet, J; Solinas, G; Montani, J; Schutz, Y. Body composition phenotypes in pathways to obesity and the metabolic syndrome. *Int J Obes*, 2010, 34, S4–17.

[40] Cruz-Jentoft, AJ; Baeyens, JP; Bauer, JM; Boirie, Y; Cederholm, T; Landi, F; et al. Sarcopenia: European consensus on definition and diagnosis. *Age Ageing*, 2010, 39, 412–23.

[41] Moran, J; Wilson, F; Guinan, E; McCormick, P; Hussey, J; Moriarty, J. Role of cardiopulmonary exercise testing as a risk-assessment method in patients undergoing intra-abdominal surgery: A systematic review. *Br J Anaesth*, 2016, 116, 177–91.

[42] Hennis, PJ; Meale, PM; Grocott, MPW. Cardiopulmonary exercise testing for the evaluation of perioperative risk in non-cardiopulmonary surgery. *Postgr Med J*, 2011, 87, 550–7.

[43] Weisman, Im. Clinical Exercise Testing. *Clin Chest Med*, 2001, 22, 366–78.
[44] Pallarés, J; Morán-Navarro, R. Methodological approach to the Cardiorespiratory Endurance Training. *J Sport Heal Res*, 2012, 4, 119–36.
[45] Piazza, O; Miccichè, V; Esposito, C; Romano, G; De Robertis, E. Individualised prediction of postoperative cardiorespiratory complications after upper abdominal surgery. *Trends Anaesth Crit Care*, 2016, 6, 11–9.
[46] Lieffers, JR; Bathe, OF; Fassbender, K; Winget, M; Baracos, VE. Sarcopenia is associated with postoperative infection and delayed recovery from colorectal cancer resection surgery. *Br J Cancer*, 2012, 107, 931–6.
[47] Huang, DD; Zhou, CJ; Wang, SL; Mao, ST; Zhou, XY; Lou, N; et al. Impact of different sarcopenia stages on the postoperative outcomes after radical gastrectomy for gastric cancer. *Surg (United States)*, 2017, 161, 680–93.
[48] Moran, J; Wilson, F; Guinan, E; McCormick, P; Hussey, J; Moriarty, J. Role of cardiopulmonary exercise testing as a risk-assessment method in patients undergoing intra-abdominal surgery: A systematic review. *Br J Anaesth*, 2016, 116, 177–91.
[49] Smith, TB; Stonell, C; Purkayastha, S; Paraskevas, P. Cardiopulmonary exercise testing as a risk assessment method in non cardio-pulmonary surgery: a systematic review. *Anaesthesia*, 2009, 64, 883–93.
[50] Lee, CHA; Kong, JC; Ismail, H; Riedel, B; Heriot, A. Systematic Review and Meta-analysis of Objective Assessment of Physical Fitness in Patients Undergoing Colorectal Cancer Surgery. *Dis Colon Rectum*, 2018, 61, 400–9.
[51] McCullough, PA; Gallagher, MJ; DeJong, AT; Sandberg, KR; Trivax, JE; Alexander, D; et al. Cardiorespiratory fitness and short-term complications after bariatric surgery. *Chest*, 2006, 130, 517–25.

[52] West, MA; Lythgoe, D; Barben, CP; Noble, L; Kemp, GJ; Jack, S; et al. Cardiopulmonary exercise variables are associated with postoperative morbidity after major colonic surgery: A prospective blinded observational study. *Br J Anaesth*, 2014, 112, 665–71.

[53] Hartley, RA; Pichel, AC; Grant, SW; Hickey, GL; Lancaster, PS; Wisely, NA; et al. Preoperative cardiopulmonary exercise testing and risk of early mortality following abdominal aortic aneurysm repair. *Br J Surg*, 2012, 99, 1539–46.

[54] Bernal, W; Martin-Mateos, R; Lipcsey, M; Tallis, C; Woodsford, K; McPhail, M; et al. Aerobic capacity during cardiopulmonary exercise testing and survival with and without liver transplantation for patients with chronic liver disease. *Liver Transplant*, 2014, 20, 54–62.

[55] Junejo, MA; Mason, JM; Sheen, AJ; Moore, J; Foster, P; Atkinson, D; et al. Cardiopulmonary exercise testing for preoperative risk assessment before hepatic resection. *Br J Surg*, 2012, 99, 1097–104.

[56] Lai, CW; Minto, G; Challand, CP; Hosie, KB; Sneyd, JR; Creanor, S; et al. Patients' inability to perform a preoperative cardiopulmonary exercise test or demonstrate an anaerobic threshold is associated with inferior outcomes after major colorectal surgery. *Br J Anaesth*, 2013, 111, 607–11.

[57] Hennis, PJ; Meale, PM; Hurst, RA; Doherty, AFO; Otto, J; Kuper, M; et al. Cardiopulmonary exercise testing predicts postoperative outcome in patients undergoing gastric bypass surgery. *Br J Anaesth*, 2012, 109, 566–71.

[58] Brunelli, A; Kim, AW; Berger, KI; Addrizzo-Harris, DJ. Physiologic evaluation of the patient with lung cancer being considered for resectional surgery: Diagnosis and management of lung cancer, 3rd ed: American college of chest physicians evidence-based clinical practice guidelines. *Chest*, 2013, 143, e166S–e190S.

[59] Older, P; Smith, R; Courtney, P; Hone, R. Preoperative evaluation of cardiac failure and ischemia in elderly patients by cardiopulmonary exercise testing. *Chest*, 1993, 104, 701–4.

[60] Giannoudis, PV; Dinopoulos, H; Chalidis, B; Hall, GM. Surgical stress response. *Injury*, 2006, 37, S3–9.

[61] Singh, M; Anaesthesiology, H; Singh, MM. Stress Response and Anaesthesia Altering the Peri and Post-Operative Management. *Indian J Anaesth*, 2003, 47, 427–34.

[62] Gillis, C; Carli, F. Promoting Perioperative Metabolic and Nutritional Care. *Anesthesiology*, 2015, 123, 1455–72.

[63] Agnew, N. Preoperative cardiopulmonary exercise testing. *Contin Educ Anaesthesia, Crit Care Pain*, 2010, 10, 33–7.

[64] Shoemaker, W; Appel, P; Kram, H. Role of oxygen debt in the development of organ failure sepsis, and death in high-risk surgical patients. *Chest*, 1992, 102, 208–15.

[65] Lutz, CT; Quinn, LBS. Sarcopenia, obesity, and natural killer cell immune senescence in aging: Altered cytokine levels as a common mechanism. *Aging (Albany NY)*, 2012, 4, 535–46.

[66] Van Venrooij, LMW; De Vos, R; Zijlstra, E; Borgmeijer-Hoelen, MMMJ; Van Leeuwen, PAM; De Mol, BAJM. The impact of low preoperative fat-free body mass on infections and length of stay after cardiac surgery: A prospective cohort study. *J Thorac Cardiovasc Surg*, 2011, 142, 1263–9.

[67] Windsor, JA; Hill, GL. Protein Depletion and Surgical Risk. *Aust N Z J Surg*, 1988, 58, 711–5.

[68] Van Venrooij, LMW; Verberne, HJ; de Vos, R; Borgmeijer-Hoelen, MMMJ; van Leeuwen, PAM; de Mol, BAJM. Postoperative loss of skeletal muscle mass, complications and quality of life in patients undergoing cardiac surgery. *Nutrition*, 2012, 28, 40–5.

[69] Nygren, J; Thorell, A; Efendic, S; Nair, KS; Ljungqvist, O. Site of insulin resistance after surgery: the contribution of hypocaloric nutrition and bed rest. *Clin Sci (Lond)*, 1997, 93, 137–46.

[70] Biteker, M; Dayan, A; Can, MM; Ilhan, E; Biteker, FS; Tekkeşin, A; et al. Impaired fasting glucose is associated with increased perioperative cardiovascular event rates in patients undergoing major non-cardiothoracic surgery. *Cardiovasc Diabetol*, 2011, 10, 1–7.

[71] Jensen, J; Rustad, PI; Kolnes, AJ; Lai, YC. The role of skeletal muscle glycogen breakdown for regulation of insulin sensitivity by exercise. *Front Physiol*, 2011, 2, 1–11.

[72] Moriello, C; Mayo, NE; Feldman, L; Carli, F. Validating the Six-Minute Walk Test as a Measure of Recovery After Elective Colon Resection Surgery. *Arch Phys Med Rehabil*, 2008, 89, 1083–9.

[73] Li, C; Carli, F; Lee, L; Charlebois, P; Stein, B; Liberman, AS; et al. Impact of a trimodal prehabilitation program on functional recovery after colorectal cancer surgery: A pilot study. *Surg Endosc Other Interv Tech*, 2013, 27, 1072–82.

[74] Kortebein, P; Ferrando, A; Lombeida, J; Wolfe, R; Evans, WJ. Effect of 10 Days of Bed Rest on Skeletal Muscle in Healthy Older Adults. *JAMA*, 2007, 297, 1772–4.

[75] Puthucheary, ZA; Rawal, J; McPhail, M; Connolly, B; Ratnayake, G; Chan, P; et al. Acute skeletal muscle wasting in critical illness. *JAMA - J Am Med Assoc*, 2013, 310, 1591–600.

[76] Pišot, R; Marusic, U; Biolo, G; Mazzucco, S; Lazzer, S; Grassi, B; et al. Greater loss in muscle mass and function but smaller metabolic alterations in older compared with younger men following 2 wk of bed rest and recovery. *J Appl Physiol*, 2016, 120, 922–9.

[77] Santa Mina, D; Scheede-Bergdahl, C; Gillis, C; Carli, F. Optimization of surgical outcomes with prehabilitation. *Appl Physiol Nutr Metab*, 2015, 40, 966–9.

[78] Zeng, C; li, H; Yang, T; Deng, ZH; Yang, Y; Zhang, Y; et al. Electrical stimulation for pain relief in knee osteoarthritis: Systematic review and network meta-analysis. *Osteoarthr Cartil*, 2015, 23, 189–202.

[79] Vance, CGT; Dailey, DL; Rakel, BA; Sluka, KA. Using TENS for pain control: the state of the evidence. *Pain Manag*, 2014, 4, 197–209.

[80] Peckham, PH; Knutson, JS. Functional Electrical Stimulation for Neuromuscular Applications. *Annu Rev Biomed Eng*, 2005, 7, 327–60.

[81] O'Connor, D; Brennan, L; Caulfield, B. The use of neuromuscular electrical stimulation (NMES) for managing the complications of ageing related to reduced exercise participation. *Maturitas*, 2018, 113, 13–20.

[82] Wernbom, M; Augustsson, J; Thomeé, R. The influence of frequency, intensity, volume and mode of strength training on whole muscle cross-sectional area in humans. *Sports Med*, 2007, 37, 225–64.

[83] Ruther, CL; Golden, CL; Harris, RT; Dudley, GA. Hypertrophy, resistance training, and the nature of skeletal muscle activation. *J Strength Cond Res*, 1995, 9, 155–9.

[84] Cabric, M; Appell, HJ. Effect of electrical stimulation of high and low frequency on maximum isometric force and some morphological characteristics in men. *Int J Sport Med*, 1987, 8, 256–60.

[85] Gondin, J; Guette, M; Ballay, Y; Martin, A. Electromyostimulation training effects on neural drive and muscle architecture. *Med Sci Sports Exerc*, 2005, 37, 1291–9.

[86] Dirks, ML; Wall, BT; Snijders, T; Ottenbros, CLP; Verdijk, LB; Van Loon, LJC. Neuromuscular electrical stimulation prevents muscle disuse atrophy during leg immobilization in humans. *Acta Physiol*, 2014, 210, 628–41.

[87] Gibson, JNA; Smith, K; Rennie, MJ. Prevention of Disuse Muscle Atrophy By Means of Electrical Stimulation: Maintenance of Protein Synthesis. *Lancet*, 1988, 332, 767–70.

[88] Gerovasili, V; Stefanidis, K; Vitzilaios, K; Karatzanos, E; Politis, P; Koroneos, A; et al. Electrical muscle stimulation preserves the

muscle mass of critically ill patients: A randomized study. *Crit Care*, 2009, 13, 1–8.

[89] Segers, J; Hermans, G; Bruyninckx, F; Meyfroidt, G; Langer, D; Gosselink, R. Feasibility of neuromuscular electrical stimulation in critically ill patients. *J Crit Care*, 2014, 29, 1082–8.

[90] Esteve, V; Carneiro, J; Moreno, F; Fulquet, M; Garriga, S; Pou, M; et al. The effect of neuromuscular electrical stimulation on muscle strength; functional capacity and body composition in haemodialysis patients. *Nefrologia*, 2017, 37, 68–77.

[91] Urdampilleta, A; Vicente-Salar, N; Martínez-Sanz, J. Necesidades proteicas de los deportistas y pautas diétetico-nutricionales para la ganancia de masa muscular. *Rev Española Nutr Humana y Dietética ...*, 2012, 16, 25–35.

[92] English, KL; Paddon-Jones, D. Protecting muscle mass and function in older adults during bed rest. *Curr Opin Clin Nutr Metab Care*, 2010, 13, 34–9.

[93] Kubiak, RJ; Whitman, KM; Johnston, RM. Changes in quadriceps femoris muscle strength using isometric exercise versus electrical stimulation. *J Orthop Sports Phys Ther*, 1987, 8, 537–41.

[94] Colson, S; Martin, A; Van Hoecke, J. Re-examination of training effects by electrostimulation in the human elbow musculoskeletal system. *Int J Sports Med*, 2000, 21, 281–8.

[95] Selkowitz, DM. Improvement in isometric strength of the quadriceps femoris muscle after training with electrical stimulation. *Phys Ther*, 1985, 65, 186–96.

[96] Porcari, JP; Miller, J; Cornwell, K; Foster, C; Gibson, M; McLean, K; et al. The effects of neuromuscular electrical stimulation training on abdominal strength, endurance, and selected anthropometric measures. *J Sport Sci Med*, 2005, 4, 66–75.

[97] Porcari, J; Ryskey, A; Foster, C. The Effects of High Intensity Neuromuscular Electrical Stimulation on Abdominal Strength and Endurance, Core Strength, Abdominal Girth, and Perceived Body Shape and Satisfaction. *Int J Kinesiol Sport Sci*, 2018, 6, 19.

[98] Hainaut, K; Duchateau, J. Neuromuscular electrical stimulation and voluntary exercise. *Sport Med*, 1992, 14, 100–13.

[99] Bax, L; Staes, F; Verhagen, A. Does neuromuscular electrical stimulation strengthen the quadriceps femoris? A systematic review of randomised controlled trials. *Sport Med*, 2005, 35, 191–212.

[100] Wageck, B; Nunes, GS; Silva, FL; Damasceno, MCP; de Noronha, M. Application and effects of neuromuscular electrical stimulation in critically ill patients: Systematic review. *Med Intensiva*, 2014, 38, 444–54.

[101] Fischer, A; Spiegl, M; Altmann, K; Winkler, A; Salamon, A; Themessl-Huber, M; et al. Muscle mass, strength and functional outcomes in critically ill patients after cardiothoracic surgery: Does neuromuscular electrical stimulation help? The Catastim 2 randomized controlled trial. *Crit Care*, 2016, 20, 1–13.

[102] Jones, S; Wdc, M; Gao, W; Ij, H; Wilcock, A; Maddocks, M. Neuromuscular electrical stimulation for muscle weakness in adults with advanced disease (Review). *Cochrane Libr*, 2016.

[103] Esteve, V; Carneiro, J; Moreno, F; Fulquet, M; Garriga, S; Pou, M; et al. The effect of neuromuscular electrical stimulation on muscle strength, functional capacity and body composition in haemodialysis patients. *Nefrologia*, 2017, 37, 68–77.

[104] Banerjee, P; Caulfield, B; Crowe, L; Clark, AL. Prolonged Electrical Muscle Stimulation Exercise Improves Strength, Peak VO2, and Exercise Capacity in Patients With Stable Chronic Heart Failure. *J Card Fail*, 2009, 15, 319–26.

[105] Deley, G; Eicher, JC; Verges, B; Wolf, JE; Casillas, JM. Do low-frequency electrical myostimulation and aerobic training similarly improve performance in chronic heart failure patients with different exercise capacities? *J Rehabil Med*, 2008, 40, 219–24.

[106] Hill, K; Cavalheri, V; Mathur, S; Roig, M; Robles, P; Te, D; et al. Neuromuscular electrostimulation for adults with chronic

obstructive pulmonary disease (Review). *Send to Cochrane Database Syst Rev*, 2018.

[107] Veldman, MP; Gondin, J; Place, N; Maffiuletti, NA. Effects of neuromuscular electrical stimulation training on endurance performance. *Front Physiol*, 2016, 7, 544.

[108] Banerjee, P; Clark, A; Witte, K; Crowe, L; Caulfield, B. Electrical stimulation of unloaded muscles causes cardiovascular exercise by increasing oxygen demand. *Eur J Cardiovasc Prev Rehabil*, 2005, 12, 503–8.

[109] Vaquero, AF; Chicharro, JL; Gil, L; Ruiz, MP; Sanchez, V; Lucia, A; et al. Effects of muscle electrical stimulation on peak VO2 in cardiac transplant patients. *Int J Sports Med*, 1998, 19, 317–22.

[110] Henneman, E. Relation between size of neuron and their susceptibility to discharge. *Science*, (80-), 1957, 126, 1345–7.

[111] Gregory, CM; Bickel, CS. Recruitment patterns in human skeletal muscle during electrical stimulation. *Phys Ther*, 2005, 85, 358–64.

[112] Bickel, CS; Gregory, CM; Dean, JC. Motor unit recruitment during neuromuscular electrical stimulation: A critical appraisal. *Eur J Appl Physiol*, 2011, 111, 2399–407.

[113] Vanderthommen, M; Depresseux, J; Dauchat, L; Degueldre, C; Croisier, J; Crielaard, J. Spatial distribution of blood flow in electrically stimulated human muscle: a positron emission tomography study. *Muscle Nerve*, 2000, 23, 482–9.

[114] Boerio, D; Jubeau, M; Zory, R; Maffiuletti, NA. Central and peripheral fatigue after electrostimulation-induced resistance exercise. *Med Sci Sports Exerc*, 2005, 37, 973–8.

[115] Jubeau, M; Sartorio, A; Marinone, PG; Agosti, F; Hoecke, JV; Nosaka, K; et al. Comparison between voluntary and stimulated contractions of the quadriceps femoris for growth hormone response and muscle damage. *J Appl Physiol*, 2007, 104, 75–81.

[116] Lake, DA. Neuromuscular electrical stimulation. An overview and its application in the treatment of sports injuries. *Sports Med*, 1992, 13, 320–36.

[117] Jennett, S. *Churchill Livingstone's dictionary of sport and exercise science and medicine*. Churchill Livingstone Elsevier, 2008.
[118] Maffiuletti, NA. Physiological and methodological considerations for the use of neuromuscular electrical stimulation. *Eur J Appl Physiol*, 2010, 110, 223–34.
[119] Adams, GR; Harris, RT; Woodard, D; Dudley, GA. Mapping of electrical muscle stimulation using MRI. *J Appl Physiol*, 1993, 74, 532–7.
[120] Herzig, D; Maffiuletti, NA; Eser, P. The Application of Neuromuscular Electrical Stimulation Training in Various Non-neurologic Patient Populations: A Narrative Review. *PMR*, 2015, 7, 1167–78.
[121] Selkowitz, DM. High frequency electrical stimulation in muscle strengthening. A review and discussion. *Am J Sport Med*, 1989, 17, 103–11.
[122] Strauss, GR. The Effect of Different Electro-Motor Stimulation Training Intensities on Strength Improvement. *Aust J Physiother*, 1988, 34, 151–64.
[123] Gregory, CM; Dixon, W; Bickel, CS. Impact of varying pulse frequency and duration on muscle torque production and fatigue. *Muscle and Nerve*, 2007, 35, 504–9.
[124] Gorgey, AS; Black, CD; Elder, CP; Dudley, GA. Effects of Electrical Stimulation Parameters on Fatigue in Skeletal Muscle. *J Orthop Sport Phys Ther*, 2009, 39, 684–92.
[125] Vivodtzev, I; Lacasse, Y; Maltais, F. Neuromuscular electrical stimulation of the lower limbs in patients with chronic obstructive pulmonary disease. *J Cardiopulm ...*, 2008, 28, 79–91.
[126] Minogue, CM; Caulfield, BM; Reilly, RB. What are the electrical stimulation design parameters for maximum VO2 aimed aimed at Cardio-Pulmonary rehabilitation? *Annu Int Conf IEEE Eng Med Biol – Proc*, 2007, 2007, 2428–31.
[127] Carty, A; McCormack, K; Coughlan, GF; Crowe, L; Caulfield, B. Increased aerobic fitness after neuromuscular electrical stimulation

training in adults with spinal cord injury. *Arch Phys Med Rehabil*, 2012, 93, 790–5.

[128] Crognale, D; De Vito, G; Grosset, JF; Crowe, L; Minogue, C; Caulfield, B. Neuromuscular electrical stimulation can elicit aerobic exercise response without undue discomfort in healthy physically active adults. *J Strength Cond Res*, 2013, 27, 208–15.

[129] Vanderthommen, M; Duchateau, J. In contrast, tetanic contractions induced by pulses of low intensity and long duration favor the normal recruitment of motor units (size principle) (25) and neural adaptations through reflex inputs to the spinal cord and supraspinal centers. *Exerc Sport Sci Rev*, 2007, 35, 180–5.

[130] Enoka, RM. Muscle strength and its development: New perspectives. *Sport Med*, 1988, 6, 146–68.

[131] Deley, G; Babault, N. Could low-frequency electromyostimulation training be an effective alternative to endurance training? An overview in one adult. *J Sport Sci Med*, 2014, 13, 444–50.

[132] McLoda, TA; Carmack, JA. Optimal Burst Duration during a Facilitated Quadriceps Femoris Contraction. *J Athl Train*, 2000, 35, 145–50.

[133] Packman-Braun, R. Relationship between functional electrical stimulation duty cycle and fatigue in wrist extensor muscles of patients with hemiparesis. *Phys Ther*, 1988, 68, 51–6.

[134] Glaviano, NR; Saliba, S. Can the Use of Neuromuscular Electrical Stimulation Be Improved to Optimize Quadriceps Strengthening? *Sports Health*, 2016, 8, 79–85.

[135] Matheson, GO; Dunlop, RJ; McKenzie, DC; Smith, CF; Allen, PS. Force output and energy metabolism during neuromuscular electrical stimulation: a 31P-NMR study. *Scand J Rehabil Med*, 1997, 29, 175–80.

[136] Hamada, T; Sasaki, H; Hayashi, T; Moritani, T; Nakao, K. Enhancement of whole body glucose uptake during and after human skeletal muscle low-frequency electrical stimulation. *J Appl Physiol*, 2003, 94, 2107–12.

[137] Binder-Macleod, Sa; Halden, EE; Jungles, Ka. Effects of stimulation intensity on the physiological responses of human motor units. *Med Sci Sports Exerc*, 1995, 27, 556–65.

[138] Maffiuletti, NA; Gondin, J; Place, N; Stevens-Lapsley, J; Vivodtzev, I; Minetto, MA. Clinical Use of Neuromuscular Electrical Stimulation for Neuromuscular Rehabilitation: What Are We Overlooking? *Arch Phys Med Rehabil*, 2018, 99, 806–12.

[139] Theurel, J; Lepers, R; Pardon, L; Maffiuletti, NA. Differences in cardiorespiratory and neuromuscular responses between voluntary and stimulated contractions of the quadriceps femoris muscle. *Respir Physiol Neurobiol*, 2007, 157, 341–7.

[140] Gorgey, AS; Mahoney, E; Kendall, T; Dudley, GA. Effects of neuromuscular electrical stimulation parameters on specific tension. *Eur J Appl Physiol*, 2006, 97, 737–44.

[141] Doucet, BM; Lam, A; Griffin, L. Neuromuscular electrical stimulation for skeletal muscle function. *Yale J Biol Med*, 2012, 85, 201–15.

[142] Bickel, CS; Gregory, CM; Azuero, A. Matching initial torque with different stimulation parameters influences skeletal muscle fatigue. *J Rehabil Res Dev*, 2012, 49, 323.

[143] Broderick, BJ; Kennedy, C; Breen, PP; Kearns, SR; Ólaighin, G. Patient tolerance of neuromuscular electrical stimulation (NMES) in the presence of orthopaedic implants. *Med Eng Phys*, 2011, 33, 56–61.

[144] Delito, A; Shulman, A; Strube, MSMS. A study of discomfort with electrical stimulation. *Phys Ther*, 1992, 72, 410–24.

[145] Gondin, J; Cozzone, PJ; Bendahan, D. Is high-frequency neuromuscular electrical stimulation a suitable tool for muscle performance improvement in both healthy humans and athletes? *Eur J Appl Physiol*, 2011, 111, 2473–87.

[146] Alon, C; Kantor, G; Ho, HS. Effects of Electrode Size on Basic Excitatory Responses and on Selected Stimulus Parameters. *J Orthop Sport Phys Ther*, 1994, 20, 29–35.

[147] Gobbo, M; Gaffurini, P; Bissolotti, L; Esposito, F; Orizio, C. Transcutaneous neuromuscular electrical stimulation, Influence of electrode positioning and stimulus amplitude settings on muscle response. *Eur J Appl Physiol*, 2011, 111, 2451–9.

[148] Gobbo, M; Maffiuletti, NA; Orizio, C; Minetto, MA. Muscle motor point identification is essential for optimizing neuromuscular electrical stimulation use. *J Neuroeng Rehabil*, 2014, 11, 1–6.

[149] Botter, A; Oprandi, G; Lanfranco, F; Allasia, S; Maffiuletti, NA; Minetto, MA. Atlas of the muscle motor points for the lower limb: Implications for electrical stimulation procedures and electrode positioning. *Eur J Appl Physiol*, 2011, 111, 2461–71.

[150] Brooks, ME; Smith, EM; Currier, D. Effect of Longitudinal Versus Transverse Electrode Placement on Torque Production by the Quadriceps Femoris Muscle during Neuromuscular Electrical Stimulation. *J Orthop Sports Phys Ther*, 1990, 11, 530–4.

[151] Franz, A; Klaas, J; Schumann, M; Frankewitsch, T; Filler, TJ; Behringer, M. Anatomical versus functional motor points of selected upper body muscles. *Muscle and Nerve*, 2018, 57, 460–5.

[152] Behringer, M; Franz, A; McCourt, M; Mester, J. Motor point map of upper body muscles. *Eur J Appl Physiol*, 2014, 114, 1605–17.

In: Uses of Electrical Stimulation… ISBN: 978-1-53615-036-0
Editor: Jaime Ruiz-Tovar © 2019 Nova Science Publishers, Inc.

Chapter 8

INTRAOPERATIVE NEUROMONITORING OF THE RECURRENT LARYNGEAL NERVE

Manuel Durán Poveda[1], Leire Zarain Obrador[2], Alejandro García Muñoz-Najar[2], Jaime Ruiz-Tovar[2]*
and Gianlorenzo Dionigi[3]
[1]Faculty of Health Sciences,
King Juan Carlos University, Madrid, Spain
Chief of Endocrine Surgery,
Head of the Department of General Surgery,
King Juan Carlos University Hospital, Madrid, Spain
[2]Endocrine and Bariatric Surgery, Department of General Surgery,
King Juan Carlos University Hospital. Madrid, Spain
[3]Division of Endocrine and Minimally Invasive Surgery,
Department of Human Pathology in Adulthood
and Childhood "G. Barresi," University Hospital "G. Martino,"
University of Messina, Messina, Italy

---

* Corresponding author: Jaime Ruiz-Tovar, MD, PhD, E-mail: jruiztovar@gmail.com.

Abstract

One of the most feared complications in thyroid surgery is injury to the recurrent laryngeal nerve (RLN). Intraoperative nerve monitoring (IONM) has been introduced with the goal of reducing the rate of RLN injury. Although its routine use remains controversial, it could potentially assist in the neural identification/neural mapping, dissection and intraoperative prediction of postoperative function (prognosis) of the RLN including intraoperative injury lesion site identification. In this chapter we will review the actual evidence about IONM and its usefulness for reducing RLN damage.

Keywords: recurrent laryngeal nerve, intraoperative nerve monitoring

Introduction

One of the most feared complications in thyroid surgery is injury to the recurrent laryngeal nerve (RLN). It is directly related to the surgeon's performance and it leads to a significant deterioration in the patient's quality of life. Unilateral vocal cord palsy can lead to morbidity related to voice changes, especially in professional voice users, as well as potential dysphagia and aspiration. Furthermore, bilateral vocal cord palsy may require tracheostomy.

There are a number of key points which must be taken into account for RLN management during thyroid surgery: *extensive knowledge of surgical anatomy of the RLN*, *cervical exposure* and *routine visual identification of RLN*, *gained experience* and *training in thyroid surgery* and the development of a *meticulous surgical technique*. It is important to perform a laryngeal exam in all patients preoperatively and postoperatively as laryngoscopy enables us to assess the function of the vocal cords. Intraoperative nerve monitoring (IONM) may improve the standards of thyroid surgery. Intraoperative diagnosis of RLN injury may be undetected by the surgeon's eye and only an electromyographic response to an electrical stimulation of the nerve (IONM) can exclude such injury.

IONM has been introduced with the goal of reducing the rate of RLN injury. Although its routine use remains controversial, it could potentially assist in the neural identification/neural mapping, dissection and intraoperative prediction of postoperative function (prognosis) of the RLN including intraoperative injury lesion site identification. Visual examination of the surgical field is vastly insufficient to identify nerve injury and prognosticate postoperative RLN function.

There are different causes that can lead to nerve injury during surgery. However, it is important to note that the main cause of postoperative RLN dysfunction is not transection but it is nonstructural functional nerve injury. We must emphasize that the systematic search of the nerve and its visual identification are mandatory while performing a thyroidectomy.

Although visual identification of the nerve has helped to decrease the rates of RLN palsy, it does not completely prevent nerve damage as unexpected RLN palsy still occurs despite anatomic preservation of the nerve. In addition, most nerve injuries are not recognized in the operative field, so visualization of the nerve alone does not allow us to determine the degree of injury.

Thyroid surgeons should be aware that anatomical indemnity of the nerve is not synonymous of functional integrity. Actually, one of the most frequent causes of nerve injury with anatomical integrity is nerve traction. It is responsible for transient RLN palsy in most cases. Other causes of intraoperative injury with anatomical nerve integrity are compression, traction, contusion, pressure, ischemia or thermal injury, and are related with surgical technique.

Intraoperative monitoring of the recurrent laryngeal nerve represents an adjunct to routine visual identification of the nerve during surgery. Although the use of this tool in clinical practice is not completely

established yet, it has become widespread in recent years for three main reasons:
- It can be useful for the nerve mapping along its cervical course and then visually identified through direct dissection. Once the nerve is identified, additional stimulation of the adjacent non-neural tissue and the nerve, will allow identifying the nerve and its branches through the surgical field, which is especially useful in revision surgery where fibrosis and scarred tissue are predominant.
- It can help in the dissection once the nerve has been identified giving permanent information in real time (when using continuous nerve stimulation) of laryngeal nerve function.
- It can aid in the prognostication of postoperative vocal cord function, with great significance in the prevention of bilateral vocal cord palsy when a total thyroidectomy is previously planned, which can modify the surgeon's decision to proceed a bilateral surgery and perform a stage thyroidectomy.

Despite the fact that these three reasons are important enough to consider the added value of IONM in thyroid surgery, the most important issue lies in the ability to predict the postoperative function of the glottis during surgery.

A recent survey suggests that 53% of general surgeons and up to 65% of otolaryngologists in the U.S. currently employ IONM in some or all of their cases. Recent studies have pointed out the benefit of nerve monitoring for younger surgeons and low volume thyroid surgeons, as it makes the nerve identification easier and therefore increases the surgeon's confidence. IONM also contributes to the resident's training as it incorporates new technology to the surgical armamentarium and provides a standardized technique of thyroidectomy with a functional nerve impact at the end of the procedure and it reduces the potential surgeon's liability as it provides greater legal support. However, the use of nerve monitoring following thyroidectomy does not exclude malpractice litigation. It is important to transmit coherent information to patients during the preoperative period regarding the use and real benefits of neuromonitoring and not overstate the benefits of this tool. The informed consent for the use of nerve monitoring has to be realistic and reflect evidence based clinical information including both ethical and legal considerations.

The implementation of IONM depends on the surgeon's decision, but it becomes especially useful when the RLN is at risk of injury, as it assists the surgeon in the identification of the nerve, especially when operating large goiters, Graves Basedow disease, node dissection of the central compartment and particularly revision surgery. IONM may be useful in the intraoperative decision-making process as it predicts nerve functional integrity which is vital for proceeding to bilateral surgery.

IONM provides a broader view of the surgical anatomy as it incorporates neurophysiologic data that allow us to gather relevant information regarding perioperative nerve function.

However, IONM is not intended as a substitute for a comprehensive knowledge of cervical surgical anatomy and an adequate surgical technique. Visual identification and routine exposure of RLN are the "gold standard" in thyroid surgery. Following Kocher´s contribution to thyroid surgery, the only way to avoid complications during thyroidectomy is the development of these basic principles: an extensive knowledge of surgical anatomy and its numerous variations in the neck, a meticulous surgical technique including a gentle handling of the adjacent structures and a rigorous hemostasis. The incorporation of IONM to these basic principles is highly recommended.

ANATOMICAL FEATURES

With IONM, the entire course of the RLN can be traced safely. Once the nerve has been mapped out, hemostasis can be achieved safely, especially at the ligament of Berry, since RLN is particularly prone to injury at that level. The nerve can be wrapped around by the two fibrous sheets of the ligament and any medial traction during dissection of the thyroid lobe can cause a direct neuroapraxic injury to the nerve. Another scenario for nerve injury in this area is the attempt to cauterize small bleeding vessels in the proximity of the ligament that can lead to a thermal injury.

Anatomical variations are related to higher rates of RLN injury. IONM helps to identify these variations, which generally cannot be predicted preoperatively. Some of these variations are: extralaryngeal branches, intertwining of branches of the RLN and the inferior thyroid artery (ITA) and the evidence of a non-recurrent laryngeal nerve.

RLN position can be altered in patients with large goiters especially when substernal or retrotrachael extension exists. In the setting of a goitrous change, the nerve can be displaced in any direction in the surgical field placing the nerve at great risk even for experienced thyroid surgeons. In this scenario, RLN position and visual identification may be more difficult. The use of IONM in these situations facilitates the recognition of the distorted RLN and preserves its function. We strongly recommend not clamping or dividing any structure during lateral dissection of the thyroid lobe until after the RLN is definitively identified and confirmed with IONM. The routine use of nerve monitoring in these situations represents an extremely useful adjunct in surgery that assists in confirmation of the nerve.

MONITORING TECHNIQUE

Since 1980, different IONM techniques have been proposed including invasive devices such as endoscopically placed intramuscular vocal cord electrodes. However, a noninvasive technique has become the most widely used for IONM of the RLN in thyroid surgery: the endotracheal tube-based surface electrodes. The main advantages of this system include ease of setup and use, as well as it noninvasive nature.

A multidisciplinary approach and standardization of monitoring technique are of paramount importance for successful nerve monitoring.

The implementation of intraoperative monitoring technology involves two components: a method of intraoperative recurrent laryngeal nerve stimulation and a method to evaluate the response of the vocal cords to stimulation. Nerve stimulation is performed by low tension stimulation of the tissues near the RLN or indirectly, by stimulation of the vagus nerve.

Nerve monitoring should be performed following standard guidelines which include recommendations for the equipment set-up, monitor, stimulating electrodes and correct positioning of the endotracheal tube, anesthesia proceedings and the definitions of EMG monitoring data (Figure 1, 2).

Nonstandardized application of monitoring techniques leads to a great variability in its use and results.

A stimulation current of 2 mA is recommended for identification of the RLN. For confirmation of the RLN identification and further intraoperative monitoring, it is recommended to use a stimulation current of 1 mA.

The evaluation of the correct functioning of the RLN can be performed by IONM, as it allows the verification of the functional integrity of the circuit that starts in the vagus nerve (VN) and ends with the contraction of the laryngeal muscles (electromyographic response).

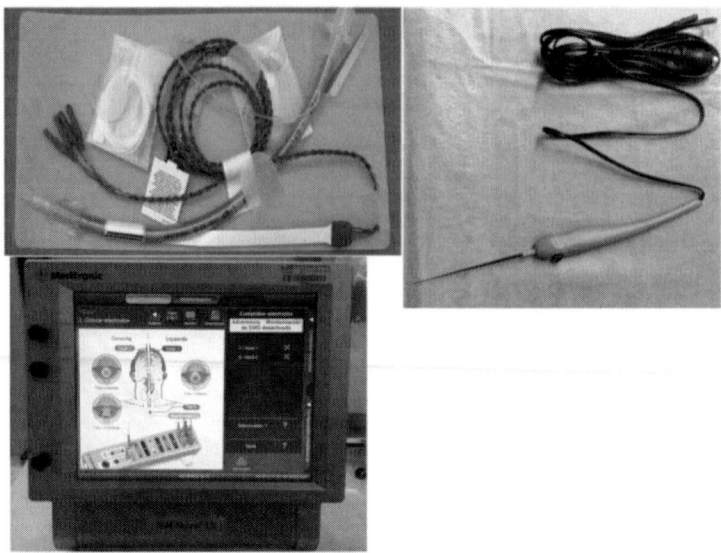

Figure 1. Endotracheal tube (NIM standard EMG reinforced tube, Medtronic Xomed) with ground electrodes; Monopolar stimulator probe (Incrementing Probe with Standard Prass Tip, Medtronic Xomed); Nerve integrity monitor (NIM-Response 3.0 System, Medtronic Xomed, Jacksonville, Florida).

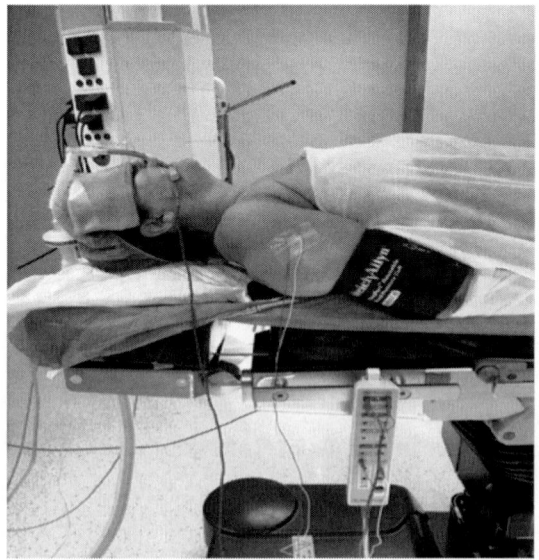

Figure 2. Adequate patient positioning with shoulder roll with head extended. Definitive tube fixation and interface connector box.

Table 1. Standardized IONM procedure

• Structured informed consent
• Preoperative laryngeal examination to assess cord mobility (L1)
• Intubation with a short acting, nondepolarizing agent. During nerve monitoring, paralyzing anesthetic agents cannot be used
• Direct visualization of monitoring tube placement after neck extension: Notice optimal depth of insertion and tube rotation
• Adequate patient positioning
• Definitive tube fixation: Secure tube in place. Repeat laryngoscopy for visualization of the glottis after patient positioning

• Equipment set-up monitor settings:
   – Respiratory variation of the baseline-waveforms amplitudes typically between 30 to 70 mV
   – Impedance values of less than 5 kΩ and impedance imbalance of less than 1 kΩ
   – Event threshold of 100 µV
   – Initial stimulation level of 1 to 2 mA

• Neural stimulation basics:
   – Vagal stimulation before thyroid dissection (V1)
   – RLN stimulation at initial identification (R1)
   – RLN stimulation at the end of thyroid dissection (R2)
   – Stimulation of EBSLN at identification (S1)
   – Stimulation of EBSLN at final dissection (S2)
Vagal stimulation after complete thyroidectomy (V2)

• Interpretation of signals during the procedure:
   – Unchanged vagal signal obtained successfully during the operation
   – LOS implies possibility of neural injury. A troubleshooting algorithm for LOS is designed to identify current problems and aid in intraoperative decisions

• EMG documentation included in patient medical charts (V1, R1, R2, V2 for each side). Waveform documentation: Amplitude, latency, waveform morphology and magnitude of stimulating current measured at the beginning, during and completion of surgery for ipsilateral RLN and vagus nerve. (Figure 3)
• Postoperative laryngeal examination (L2): Vocal cord mobility

A standardized technique of IONM includes stimulation of the VN and RLN before, during, and after thyroid gland resection.

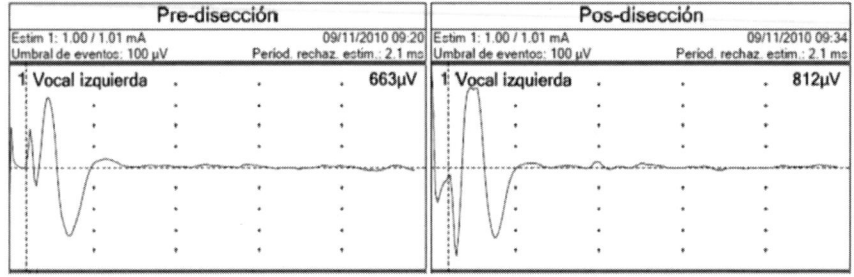

Figure 3. Left RLN recorded and documented EMG signal (R1, R2).

The minimum information which is necessary for proper IONM includes the following steps:

- Preoperative laryngeal exam (L1; preoperative laryngoscopy)
- Initial intraoperative stimulation of the VN (V1; pre-dissection vagal stimulation)
- Initial intraoperative stimulation of the RLN (R1; RLN stimulation at initial identification)
- Initial intraoperative stimulation of the external branch of the superior laryngeal nerve-EBSLN (S1; EBSLN stimulation at initial identification)
- Final stimulation of the external branch of the superior laryngeal nerve-EBSLN (S2; EBSLN stimulation at the end of dissection)

- Final intraoperative stimulation of the RLN (R2; RLN stimulation at the end of thyroid dissection)
- Final intraoperative stimulation of the VN (V2; post-dissection vagal stimulation)
- Postoperative laryngeal exam (L2; postoperative laryngoscopy)

This process finishes with the collection of all the intraoperative information in a report which must specify the EMG activity of the intervention (amplitude, latency, waveform morphology and magnitude of stimulation current of V1, R1, R2, V2 for each thyroid lobe/side) Table 1.

MONITORING EFFECTIVENESS

Apart from hypocalcemia and hematoma, recurrent laryngeal nerve paralysis (RLNP) is the most serious complication after a thyroidectomy. Laryngeal nerve injury is a severe encountered complication of thyroid surgery and it is a source of important morbidity.

Symptomatic RLNP vary greatly over time and among patients but has been proved to be a major cause of impaired quality of life and has a negative impact on job performance.

IONM is not intended to substitute routine visual identification of the nerve during surgery and an adequate surgical technique. Nerve monitoring has provided to the surgical armamentarium clinical, neurophysiologic and functional patterns to surgical practice. This broader vision of surgical anatomy and technique warranted by the use of nerve monitoring affects directly the quality of surgical procedures while incorporating a functional dynamic adjunct to the traditional surgical technique.

Although previously mentioned, IONM application has proved to be useful in the following situations:

1. Identification of the RLN. The nerve is located in the surgical field through stimulation (neural mapping, Figure 4) and then visually identified through directed dissection. Therefore, nerve monitoring can facilitate anatomic identification and dissection in order to avoid iatrogenic injuries. IONM becomes especially useful in patients with anatomic variants, as well as in large intrathoracic goiters and revision surgery. RLN monitoring is a more reliable tool to assess nerve injury than visual inspection alone.
2. Aid in dissection. Once the nerve is identified, additional stimulation of the dissected field allows tracing the nerve and all its branches, so that we can differentiate the nerve from nonneural adjacent tissue. This is especially relevant and useful for the resident and fellow during its medical education and training period.
3. Prognostication of postoperative vocal cord function and lesion site identification.

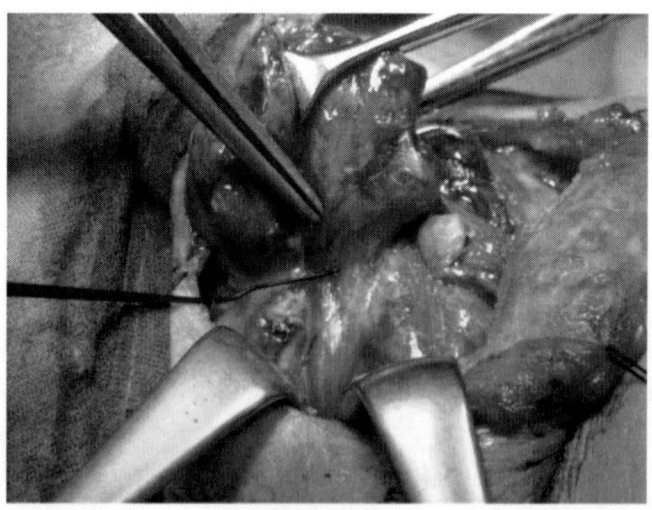

Figure 4. Neural mapping of the RLN (From Ferrero Herrero E, Durán Poveda M. Cirugía Tiroidea en Imágenes. Madrid: Editorial Dykinson. 2017).

Table 2. Summary of the impact of nerve monitoring in surgical practice

- Reduction in transient RLN palsy rate in high-risk procedures
- Intraoperative diagnosis of RLN injury
- Pronostication of postoperative neural function
- Neural identification-confirmation/neural mapping
- Aid in dissection
- Extralaryngeal RLN branching detection (Figure 5)
- Identification of functional variability of RLN branches
- Management of distorted RLN
- Management of intertwining between branches of the RLN and ITA
- Non-RLN assessment
- Aid in completeness of a total thyroidectomy
- Tumor invasion of the RLN (management of infiltrated nerve)
- Intraoperative decision-making (staged thyroidectomy)
- Useful in endoscopic thyroidectomy approaches
- Pronostication of postoperative neural function and lesion site identification
- Research
- Educational asset
- Medicolegal issues

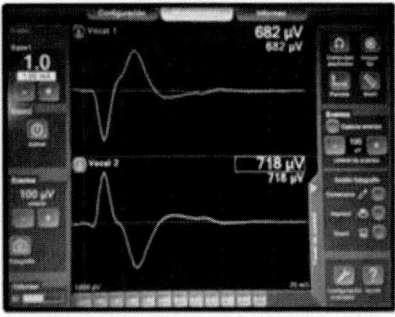
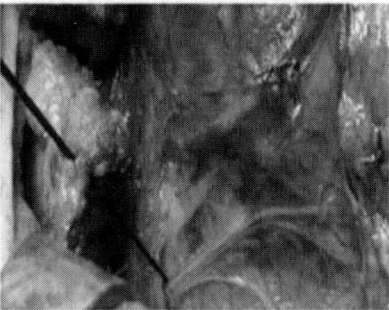

Figure 5. Identification of the anterior branch of the RLN and its EMG response with IIONM (From Ferrero Herrero E, Durán Poveda M. Cirugía Tiroidea en Imágenes. Madrid: Editorial Dykinson. 2017).

If the injury is caused by a clip, a ligature or binding, the surgery can release the cause of injury and potentially allow early recovery prior to irreversible damage.

Compression, crushing, thermal injury, ischemia, ligature, stretch or traction can block the neural conduction (neuroapraxia) without anatomical interruption of the RLN. This is why IONM represents a significant improvement in accuracy of neural function prognostic testing when compared to visual inspection of the nerve. Actually only 10-15% of nerve injuries are detected by surgeons through visual inspection. The most important aspect of monitoring is to predict the postoperative function of the vocal cords, as it allows the surgeon to avoid the morbidity related to a potential bilateral vocal cord paralysis while performing a total thyroidectomy. Therefore, the use of IONM may facilitate intraoperative decision-making for bilateral thyroid surgery so the surgeon can consider staging the procedure if loss of signal occurs.

The impact of nerve monitoring in surgical practice is summarized in Table 2.

INTERMITTENT AND CONTINUOUS INTRAOPERATIVE NEURAL MONITORING

When referring to neuromonitoring, we must point out that there are two types of IONM: intermittent and continuous monitoring.

On the one hand, intermittent IONM (IIONM) consists of the intermittent stimulation of the RLN and the vagus nerve, and therefore the intermittent registration of the nerve function whenever it is stimulated by the surgical team. The limitation of this technique is the exposure of the RLN to the risk of injury in-between two stimulations.

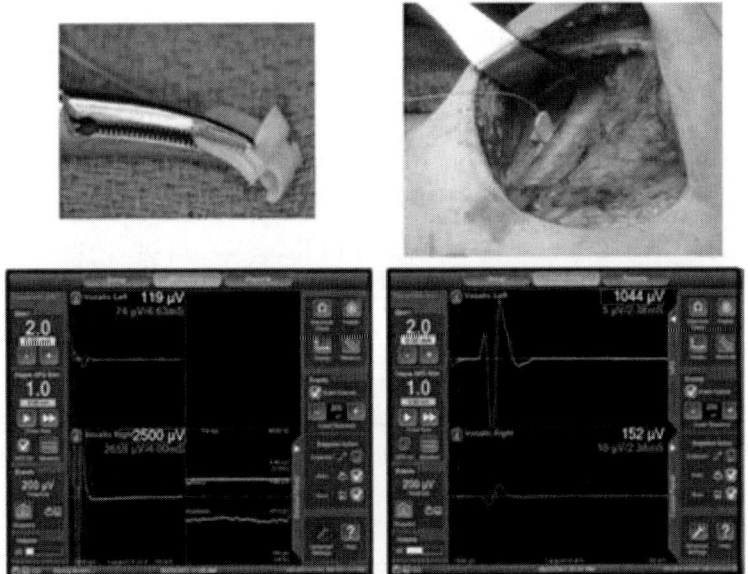

Figure 6. APS electrode used in CIONM. APS in vagus nerve. Normal EMG activity with CIONM (Courtesy of Gregory W. Randolph).

On the other hand, continuous IONM (CIONM) uses a temporary implantable vagus electrode which allows the surgical team to register the RLN function constantly (permanent surveillance of nerve function during the whole procedure) (Figure 6). This monitoring mode enables the recognition of a potentially imminent nerve injury as well as the record of intraoperative recovery of nerve function after loss of electromyographic (EMG) signal. CIONM provides in real-time instant alerts as soon as a surgical procedure impinges on the RLN. This constant feedback to the surgeon regarding nerve function makes the difference over intermittent nerve monitoring, and allows the immediate modification of certain harmful surgical maneuvers in case of adverse EMG changes during the procedure, giving the nerve a chance to recover before the damage becomes irreversible.

Therefore, CIONM has been developed to overcome the main limitation when intermittent stimulation of the RLN is used. Nerve malfunction is only detected with intermittent nerve stimulation when the nerve insult has taken place, while CIONM allows permanent evaluation of the RLN, prompt recognition of impending nerve injury and appreciates intraoperative functional nerve recovery after loss of EMG signal. Surgeon's response to EMG intraoperative changes makes the difference between IIONM and CIONM.

The use of intermittent stimulation of the RLN and the vagus nerve has now become a standard practice in numerous specialized centers. However, continuous monitoring is still in the process of being implemented.

Both IIONM and CIONM have proved to be useful during the identification of the nerve, the mapping of its course and during thyroid dissection. They are also a tool that makes it possible to diagnose segmental (type 1) and diffuse (type 2) loss of signal (LOS).

LIMITATIONS OF IONM

The most widespread modality of monitoring is the intermittent one. Unfortunately, this modality only allows the evaluation of the nerve functional integrity during the nerve stimulation period, which means that the nerve might be at risk of injury between the stimulation periods.

However, continuous monitoring alerts the surgeon in advance of certain situations that may put the nerve at risk of injury, as it provides permanent stimulation of the VN by placing a contact electrode on it. As a result, it is possible to monitor the functional integrity of the nerve along its entire length in real time and during the whole surgical procedure.

As a result, the surgeon can prevent definitive injury to the nerve by modifying some surgical maneuvers that could damage the nerve.

Continuous monitoring is still in a development stage, although many groups have already implemented it. Further investigation in this field will contribute to the correct interpretation of the wide range of information provided by c IONM, as well as to its expansion.

There are other limitations related to IONM, such as the lack of uniformity in the application of the standardized IONM technique.

We can also note that IONM increases operative time. Finally, cost-effectiveness has to be fully developed.

Table 3. C-IONM & I-IONM: complementary devices. (Modified from: Sun H, Zanghì GN, Freni F, Dionigi G. Continuous and intermitted nerve monitoring in thyroid surgery: two complementary devices. Gland Surg. 2018; 7: S80-S81)

	I-IONM	C-IONM
Prevent Transient RLNP	-	-
Prevent Permanent RLNP	-	++
RLN identification and mapping	+++	+
Nerve dissection	++	-
Injury lesión site identification	++	+
Nerve variability/ extralaryngeal branching	++	+
EBSLN monitoring	++	-
Impending loss of EMG signal/ Perceive imminent RLN stress	+	+++
Loss of the EMG signal	+++	+++
Syncronization of surgical manouvers	-	+++
Surveillance of nerve function during surgical procedure	+	+++
Recovery after loss of the EMG signal/ recognition of intraoperative recovery of nerve function after LOS	+/-	+++
Elucidates mechanism of nerve injury	+	++
EMG tube malpositioning	++	-
Documentation	+	+++
Cost	++	+++

IONM should not replace clinical judgment. We believe that IIONM should be the initial approach to nerve monitoring and with experience and hard work, move on to C-IONM. CIONM has a different learning curve compared to the intermittent modality. CIONM gives constant intraoperative information regarding nerve function and the surgeon has to be permanently aware of it. Both devices represent a surgical adjunct to conventional thyroidectomy in a complementary but not exclusive way (Table 3).

References

Angelos P. Identification of the External Branch of the Superior Laryngeal Nerve: An Additional Argument for Neuromonitoring? *Ann Surg Oncol.* 2015; 22(6):1751-2.

Barczynski M, Konturek A, Pragacz K, et al. Intraoperative Nerve Monitoring Can Reduce Prevalence of Recurrent Laryngeal Nerve Injury in Thyroid Reoperations: Results of a Retrospective Cohort Study. *World J Surg.* 2014; 38:599-606.

Barczyński M, Randolph GW, Cernea CR, et al. External branch of the superior laryngeal nerve monitoring during thyroid and parathyroid surgery: International Neural Monitoring Study Group standards guideline statement. *Laryngoscope.* 2013; 123 Suppl 4: S1-14.

Dionigi G, Bacuzzi A, Barczynski M, et al. Implementation of systematic neuromonitoring training for thyroid surgery. *Updates Surg.* 2011; 63(3):201-7.

Dionigi G, Barczynski M, Chiang FY et al. Why monitor the recurrent laryngeal nerve in thyroid surgery? *J Endocrinol Invest.* 2010; 33(11):819-22.

Dionigi G, Chiang FY, Dralle H, et al. Safety of neural monitoring in thyroid surgery. *Int J Surg.* 2013; 11 Suppl 1: S120-6.

Dionigi G, Wu CW, Kim HY, et al. Severity of Recurrent Laryngeal Nerve Injuries in Thyroid Surgery. *World J Surg.* 2016; 40(6):1373-81.

Dralle H, Sekulla C, Haerting J, et al. Risk factors of paralysis and functional outcome after recurrent laryngeal nerve monitoring in thyroid surgery. *Surgery.* 2004; 136(6):1310-22.

Durán Poveda MC, Dionigi G, Sitges-Serra A, et al. Intraoperative Monitoring of the Recurrent Laryngeal Nerve during Thyroidectomy: A Standardized Approach (Part 1). *WJOES*. 2011; 3(3): 144-150.

Durán Poveda MC, Dionigi G, Sitges-Serra A, et al. Intraoperative Monitoring of the Recurrent Laryngeal Nerve during Thyroidectomy: A Standardized Approach (Part 2). *WJOES*. 2012; 4(1): 33-40.

Durán Poveda M, García Muñoz-Najar A, Franco Herrera R et al. Aspectos Generales de la Monitorización Nerviosa Intraoperatoria del Nervio Laríngeo Recurrente en Cirugía Tiroidea. In: Ferrero Herrero E, Durán Poveda M, editors. *Cirugía Tiroidea en Imágenes*. ed. Madrid: Dykinson; 2018. p. 43-55.

Engelsman AF, Warhurst S, Fraser S, et al. Influence of neural monitoring during thyroid surgery on nerve integrity and postoperative vocal function. *BJS Open*. 2018; 25; 2(3):135-141.

Ferrero Herrero E, Durán Poveda M. *Cirugía Tiroidea en Imágenes*. Madrid: Editorial Dykinson. 2017.

Lin HS, Terris DJ. An Update on the Status of nerve monitoring for thyroid/parathyroid surgery. *Curr Opin Oncol*. 2017, 29(1): 14-19.

Périé S, Santini J, Kim HY, et al. International consensus (ICON) on comprehensive management of the laryngeal nerves risks during thyroid surgery. *Eur Ann Otorhinolaryngol Head Neck Dis*. 2018;135(1S): S7-S10.

Randolph GW, Dralle H et al. Electrophysiologic Recurrent Laryngeal Nerve Monitoring During Thyroid and Parathyroid Surgery: International Standards Guideline Statement. *Laryngoscope*. 2011; 121 Suppl 1: S1-16.

Randolph GW, Kamani D. Intraoperative Electrophysiologic Monitoring of the Recurrent Laryngeal Nerve During Thyroid and Parathyroid Surgery: Experience With 1381 Nerves at Risk. *Laryngoscope*. 2017; 127(1):280-286.

Schneider R, Randolph GW, Barczinsky M, et al. Continuous intraoperative neural monitoring of the recurrent nerves in thyroid surgery: a quantum leap in technology. *Gland Surg* 2016; 5 (6): 607-616.

Schneider R, Randolph GW, Dionigi G, et al. International neural monitoring study group guideline 2018 part I: Staging bilateral thyroid surgery with monitoring loss of signal. *Laryngoscope*. 2018; 5. doi: 10.1002/lary.27359.

Snyder SK, Lairmore TC, Hendricks JC, et al. Elucidating Mechanisms of Recurrent Laryngeal Nerve Injury During Thyroidectomy and Parathyroidectomy. *J Am Coll Surg*. 2008; 206(1):123-30.

Sun H, Zanghì GN, Freni F, Dionigi G. Continuous and intermitted nerve monitoring in thyroid surgery: two complementary devices. *Gland Surg*. 2018;7: S80-S81.

Wu CW, Dionigi G, Barczynski M, et al. International neuromonitoring study group guidelines 2018: Part II: Optimal recurrent laryngeal nerve management for invasive thyroid cancer-incorporation of surgical, laryngeal, and neural electrophysiologic data. *Laryngoscope*. 2018; 6. doi: 10.1002/lary.27360.

ABOUT THE EDITOR

Jaime Ruiz-Tovar
Department of Bariatric Surgery.
Centro de Excelencia para el Diagnóstico y
Tratamiento de la Obesidad (Valladolid-Spain)
jruiztovar@gmail.com

Prof. Jaime Ruiz-Tovar is head of the Neurostimulation Unit at Garcilaso Clinic (Madrid, Spain); Bariatric Surgeon at Centro de Excelencia para el Tratamiento de la Obesidad (Valladolid, Spain) and Rey Juan Carlos University Hospital (Madrid, Spain). He is also Professor of Surgery, Universidad Alfonso X (Madrid, Spain) and Head of the ERAS-Spain group in Bariatric Surgery

INDEX

A

acetylcholine, 3, 48, 64
acupuncture, 5, 16, 77
adenosine, 3, 8, 70, 75
adipose tissue, 35, 51
adverse effects, 20
aerobic exercise, 121
afferent nerve fiber, 7
anatomy, 46, 126, 129, 135
aneurysm, 86, 113
appendectomy, 72, 76
appendicitis, 72
appetite reduction, 31, 32, 33, 34, 36, 37, 38, 40, 41, 43, 44, 46, 50, 52
articulation, 7
aspartate, 9, 14
atrophy, 116
autonomic nervous system, 47, 66
axons, 104

B

back pain, 5, 12
biofeedback, 18, 19, 21
bleeding, 26, 62, 130
blood, 10, 11, 21, 40, 41, 43, 61, 88, 89, 94, 119
blood flow, 11, 21, 119
blood plasma, 10
blood pressure, 88, 94
bloodstream, 40
body composition, 83, 87, 92, 117, 118
body mass index (BMI), 32, 37, 38, 41, 49, 82, 83, 111
bowel, ix, 24, 56, 58, 63, 64, 65, 67
bowel obstruction, 58
bowel sounds, 56, 63
breastfeeding, 58

C

caffeine, 14, 70
caloric intake, 41
caloric restriction, 38
cancer, 73, 82, 108, 109, 112, 113
carbohydrates, 39
cardiac output, 88
cardiac surgery, 89, 114
catabolism, 70, 94

central nervous system, 33, 35, 52
cholecystectomy, 71
cholesterol, 41
chronic heart failure, 118
chronic obstructive pulmonary disease, 3, 119, 120
clinical judgment, 141
clinical trials, 69
colectomy, 61
colorectal cancer, 55, 58, 77, 85, 107, 110, 112, 115
colorectal surgeon, 21
colorectal surgery, 17, 55, 56, 63, 65, 66, 67, 80, 86, 90, 113
comorbidity, 18
complications, 20, 26, 59, 60, 63, 79, 80, 81, 82, 83, 84, 85, 86, 87, 88, 89, 90, 93, 106, 108, 109, 110, 112, 114, 116, 126, 129
constipation, 18
controlled trials, 65, 118
cortex, 13
cortisol, 51

D

data analysis, 25
decision-making process, 129
defecation, 63
dermatome, vii, 31, 32, 33, 35, 36, 37, 39, 40, 46, 47, 48, 49, 50, 51, 52
dermatome T6, vii, 31, 32, 33, 35, 36, 37, 39, 40, 46, 50, 51, 52
dermatome T7, vii, 47, 48, 49, 50, 51, 52
diabetes mellitus, vii, 32, 47, 48, 52, 53, 82, 111
diabetic neuropathy, 11, 15
dietary compliance, 32, 39
dietary habits, 18
discomfort, 97, 98, 100, 102, 103, 104, 121, 122

dizziness, 73
dorsal horn, 3, 5, 10, 11, 14, 15, 70
drug interaction, 5
drug therapy, 5
drugs, x
dyslipidemia, 32, 44
dysphagia, 126

E

electrical nerve stimulation, 2, 4, 12, 13, 14, 15, 23, 56, 57, 69, 75, 76, 77, 92
electroacupuncture, 69, 71, 75, 76
electrodes, 4, 23, 59, 70, 92, 97, 102, 103, 104, 107, 131, 132
electrotherapy, 1, 2, 3, 5, 12, 15
EMG, 131, 132, 133, 134, 135, 137, 139, 140, 141
emotional state, 72, 73
endocrine, 48, 69, 74
endogenous synthesis, 51
endorphins, 15
end-stage renal disease, 3
energy, 46, 85, 88, 121
ENS, 4, 70, 73, 74
epinephrine, 51
exercise, 80, 84, 85, 90, 92, 95, 96, 97, 111, 112, 113, 114, 115, 116, 117, 118, 119, 120
exercise participation, 116

F

faecal incontinence, 25, 28, 29, 46
fasting, 33, 40, 41, 89, 115
fasting glucose, 89, 115
fat, 83, 89, 114
fecal, vii, 3, 4, 17, 18, 19, 20, 27, 28, 34, 35, 36, 37, 45, 49, 50, 52
feces, 26

fibers, 1, 5, 6, 7, 20, 35, 96, 97, 98, 100, 102, 103, 104
fibrosis, 128
fish, 2, 3
fitness, 80, 82, 84, 85, 86, 87, 88, 90, 91, 92, 95, 100, 101, 102, 105, 106, 107, 110, 112, 120
flatulence, 56, 59, 62
food intake, 33, 37, 39

G

gastrectomy, 46, 80, 108, 112
gastrointestinal bleeding, 63
general surgeon, 128
ghrelin, 31, 32, 33, 35, 40, 41, 42, 43, 46, 51, 53
glottis, 128, 133
glucose, 41, 42, 46, 51, 53, 83, 89, 121
glutamate, 9, 14
glycogen, 89, 115
glycolysis, 85
gout, 2, 3
growth hormone, 41, 51, 119

H

health care, 27
health effects, 53
health risks, 83
health status, 25, 83
heart disease, 32
heart rate, 76, 88, 94
hematoma, 26, 135
hemiparesis, 121
hemoglobin, 41, 42, 50
hemorrhoidectomy, 71, 76
hemostasis, 129, 130
hepatic transplant, 86
hernia repair, 70
herniorrhaphy, 75

high density lipoprotein, 41
homeostasis, 51, 88
homeostatic model assessment (HOMA), 48, 50, 51
hormones, 33, 40, 51
hospital acquired pneumonia, 109
human, 13, 15, 45, 88, 89, 90, 117, 119, 121, 122
human body, 88, 89, 90
hyperglycemia, 89
hyperinsulinism, 51
hyperlipidemia, 44
hypertension, 32, 44, 82
hypertrophy, ix, 93, 98
hypoglycemia, 53
hypothalamus, 40

I

implants, 122
incontinence, vii, 3, 4, 17, 18, 19, 20, 25, 27, 28, 29, 34, 35, 36, 37, 45, 46, 49, 50, 52
infection, 58, 62, 85, 89, 112
inflammation, 9, 14
inflammatory bowel disease, 18, 46
inguinal hernia, 70, 75
insulin, 41, 47, 48, 50, 51, **52**, **53**, 114, 115
insulin resistance, 48, 50, 51, **52**, 53, 114
insulin sensitivity, 53, 115
intensive care unit, 87, 109
intervention, 20, 36, 40, 50, **56**, 57, 58, 93, 99, 135
intraoperative nerve monitoring, 126, 142
ipsilateral, 133
ischemia, 15, 87, 114, 127, **138**

L

laparoscopic cholecystectomy, 71, 76
laparoscopic surgery, 61, 64, 73, 77

laparoscopy, 81
laparotomy, 61
laryngoscopy, 126, 133, 134, 135
lean body mass, 91
life expectancy, 83
life quality, 25
ligament, 130
lipolysis, 53
liver disease, 113
liver transplant, 89, 113
liver transplantation, 113
lung cancer, 113

M

mechanism, vii, 1, 2, 3, 5, 6, 7, 8, 9, 10, 20, 33, 34, 40, 51, 65, 114, 141
medulla, 11, 14
mellitus, 32, 48, 52, 82
meta-analysis, 28, 45, 65, 67, 95, 107, 108, 111, 115
metabolic change, 88
metabolic syndrome, 44, 52, 111
metabolism, 50, 53, 121
metastasis, 5, 110
metatarsal, 22
microcirculation, 11, 15
midbrain, 10
morbidity, 32, 56, 62, 63, 65, 86, 87, 109, 113, 126, 135, 138
morphine, 71
morphology, 133, 135
mortality, 32, 62, 63, 81, 82, 83, 86, 87, 88, 89, 107, 110, 113
mortality rate, 32, 81, 87, 89
mortality risk, 32, 83
motivation, 32, 39
motor fiber, 104
mucosa, 21
muscarinic receptor, 10, 14
muscle contraction, 99, 100
muscle mass, 80, 83, 85, 89, 90, 92, 93, 94, 101, 106, 110, 115, 117
muscle performance, 122
muscle strength, 80, 83, 89, 91, 94, 95, 99, 106, 117, 118, 120
muscular mass, 80
musculoskeletal system, 117
myocardial infarction, 81

N

nasogastric tube, 59, 64
National Health and Nutrition Examination Survey, 44
natural killer cell, 114
nausea, 56, 72, 73, 74, 76
nerve, vii, ix, 1, 2, 3, 4, 5, 6, 7, 8, 9, 10, 11, 12, 13, 14, 15, 16, 17, 19, 20, 21, 22, 23, 27, 28, 29, 33, 34, 35, 36, 37, 39, 40, 41, 45, 46, 48, 50, 52, 56, 57, 58, 66, 69, 70, 71, 72, 73, 74, 75, 76, 77, 92, 100, 116, 126, 127, 128, 129, 130, 131, 133, 134, 135, 136, 137, 138, 139, 140, 141, 142, 143, 144
nerve fibers, 4, 70, 100
nervous system, 12, 15, 48, 51
neural function, 137, 138
neurohormonal, 39
neuromuscular electrical stimulation (NMES), 11, 80, 92, 93, 94, 95, 96, 97, 98, 101, 102, 103, 105, 106, 116, 117, 118, 119, 120, 121, 122, 123
neuropathic pain, 13, 75
neuropathy, 5, 11, 15
NK cells, 89

O

obesity, 3, 32, 33, 37, 44, 45, 46, 48, 51, 53, 111, 114
opioids, 3, 9, 11, 14, 75

organism, 83, 88, 92, 94, 102
osteoarthritis, 115
oxygen, 84, 88, 99, 114, 119
oxygen consumption, 88, 89, 99

P

pacemaker, 33, 35, 39
pain, viii, ix, 2, 3, 4, 5, 6, 7, 8, 12, 13, 14, 15, 16, 20, 26, 34, 36, 45, 49, 56, 57, 66, 69, 70, 71, 72, 73, 74, 75, 76, 77, 92, 97, 102, 103, 104, 114, 115, 116
pain management, 4, 66, 69, 70, 76
pancreas, ix, 47, 48, 50, 51
paralysis, 135, 138, 142
paralytic ileus, 66
parasympathetic activity, 57, 63
parathyroid, 142, 143
paresthesias, 26
pathophysiology, 66
pelvic floor, 18, 20
percutaneous electrical neurostimulation (PENS), vii, 1, 2, 3, 4, 5, 17, 31, 32, 33, 34, 35, 36, 37, 39, 40, 41, 42, 43, 46, 47, 48, 49, 50, 51
peripheral fatigue, 119
peripheral neuropathy, 3
peristalsis, 57
peritonitis, 72
phalanx, 22
pharmacological treatment, 2
phenotypes, 111
physical activity, 34, 84, 110
physical exercise, 32
physical fitness, ix, 80, 82, 85, 87, 91, 92, 93, 96, 106, 110
physical inactivity, 82
physicians, 2, 113
physiological mechanisms, 1
physiology, ix

placebo, 8, 36, 40, 65, 70, 71, 72, 73, 74, 75, 76
plasma levels, 31, 43
plasticity, 15
pneumonia, 59, 87, 89
positron emission tomography, 119
postoperative complications, 59, 80, 82, 83, 84, 85, 86, 87, 88, 89, 106, 108
postoperative ileus, 3, 55, 56, 57, 65, 66, 67
postoperative outcome, 112, 113
pregnancy, 58
prehabilitation, 80, 90, 115
premature death, 83, 111
principles, 129
prognosis, 126, 127
protein synthesis, 88, 90
psychiatric disorder, 58

Q

quadriceps, 93, 94, 97, 117, 118, 119, 122
quality of life, 17, 18, 25, 32, 69, 73, 83, 114, 126, 135

R

receptors, 2, 7, 8, 9, 10, 13, 14, 33, 40, 64, 70, 75
recovery, 63, 64, 65, 67, 71, 76, 77, 80, 81, 82, 90, 91, 92, 95, 106, 112, 115, 138, 139, 140, 141
rectus femoris, 91
recurrent laryngeal nerve, viii, ix, 125, 126, 127, 130, 131, 135, 142, 143, 144
reflexes, ix, 63
rehabilitation, 92, 94, 95, 96, 120
resection, 56, 57, 61, 65, 66, 77, 90, 109, 110, 112, 113, 134
respiratory rate, 88
risk assessment, 112, 113

S

sarcopenia, 3, 82, 83, 85, 86, 89, 112
sciatica, 5
sensation, 35, 38, 39, 40, 42, 49, 51, 98, 103, 104
sepsis, 114
serotonin, 3, 9, 11, 14, 64
skeletal muscle, 83, 89, 92, 99, 111, 114, 115, 116, 119, 121, 122
skin, 4, 5, 6, 15, 21, 51, 71, 92, 104
sleep apnea, 32
smoking, 82, 109
sphincter, 19, 34
spinal cord, 3, 6, 7, 9, 10, 14, 34, 35, 45, 51, 121
spinal cord injury, 121
standard deviation, 60
standardization, 131
stimulation, vii, ix, 1, 2, 3, 4, 5, 6, 7, 8, 9, 10, 11, 12, 13, 14, 15, 16, 19, 20, 21, 22, 23, 26, 27, 28, 29, 31, 33, 34, 35, 39, 40, 41, 43, 45, 46, 51, 52, 53, 56, 57, 58, 60, 61, 62, 63, 64, 65, 66, 69, 70, 71, 72, 73, 74, 75, 76, 77, 80, 92, 93, 95, 96, 100, 101, 102, 106, 107, 115, 116, 117, 118, 119, 120, 121, 122, 123, 127, 128, 131, 133, 134, 135, 136, 138, 139, 140
stimulus, 12, 22, 34, 123
stomach, ix, 33, 34, 35, 39
strength training, 116
stress, 88, 89, 90, 92, 114, 141
stress response, 114
substrates, 88
success rate, 20, 25
surgical intervention, 96
surgical technique, 62, 63, 79, 126, 127, 129, 135
surgical techniques, 62, 63, 79
sympathetic nervous system, 47, 48, 51
sympathetic system, 57, 63
synaptic transmission, 5

T

technological advances, 81
terminals, 35, 39
testing, 84, 85, 111, 112, 113, 114, 138
thyroid, ix, 126, 128, 129, 130, 131, 133, 134, 135, 138, 140, 141, 142, 143, 144
thyroid cancer, 144
thyroid gland, 134
tibia, 21
tibial, vii, 4, 17, 18, 19, 20, 21, 22, 23, 27, 28, 29, 33, 34, 36, 37, 45, 46, 50, 52
tissue, 50, 70, 88, 90, 128, 136
total cholesterol, 41
tracheostomy, 126
training, 92, 93, 97, 102, 105, 116, 117, 118, 119, 121, 126, 128, 136, 142
transcutaneous electrical acupoint stimulation, 69, 76, 77
transcutaneous electrical nerve stimulation (TENS), vii, 1, 2, 3, 4, 5, 6, 7, 8, 9, 10, 11, 13, 14, 16, 23, 40, 41, 57, 58, 66, 70, 71, 72, 73, 74, 75, 76, 77, 116
transection, 127
transplant, 80, 96, 119
transversus abdominis, 74, 77
trapezius, 73
type 2 diabetes, 32, 48, 52

U

urinary retention, 71
urinary tract, 46, 59
urinary tract infection, 59

V

vagus nerve, 57, 131, 133, 138, 139, 140
vasodilation, 11, 70
visual analogue scale, 69, 70
visualization, 23, 127, 133
vomiting, 56, 69, 72, 73, 74, 76

W

weight loss, 31, 32, 33, 34, 36, 37, 38, 40, 42, 45, 46, 48, 50, 51, 52, 53
weight management, 46
weight reduction, 37
weight status, 38
wells, 73

ADVANCES IN EXPERIMENTAL SURGERY. VOLUME 1

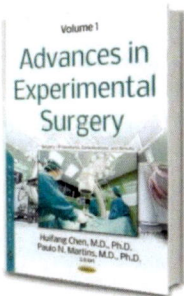

EDITORS: Huifang Chen, M.D., Ph.D. and Paulo N. Martins, M.D., Ph.D.

SERIES: Surgery – Procedures, Complications, and Results

BOOK DESCRIPTION: Volume I introduces surgical basic notions, techniques, and different surgical models involved in basic experimental surgery and review the biomechanical models, ischemia/reperfusion injury models, repair and regeneration models, and organ and tissue transplantation models, respectively.

HARDCOVER ISBN: 978-1-53612-775-1
RETAIL PRICE: $310

PERCUTANEOUS ENDOSCOPIC GASTROSTOMY (PEG): TECHNIQUES, EFFECTIVENESS AND POTENTIAL COMPLICATIONS

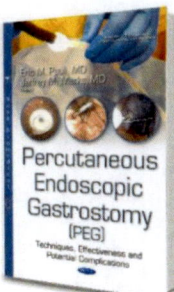

EDITORS: Eric M. Pauli, MD and Jeffrey M. Marks

SERIES: Surgery – Procedures, Complications, and Results

BOOK DESCRIPTION: The purpose of this book is twofold. First, for the individual who is new to performing PEG, it provides an overview of the basic management strategies and endoscopic techniques as they are performed by recognized experts in their respective fields. Second, for the practitioner who already performs PEG, the editor hope that it exposes them to alternative techniques and more advanced therapeutic options such that they can offer the procedure to more patients in an even safer fashion.

HARDCOVER ISBN: 978-1-63485-606-5
RETAIL PRICE: $190

Fundamentals of Leadership for Healthcare Professionals

Editors: Stanislaw P. A. Stawicki, M.D. and Michael S. Firstenberg, M.D.

Series: Health Care in Transition

Book Description: Each chapter in this text explores different aspects of healthcare leadership, provides valuable insights into how effective leadership functions, and offers practical perspectives on implementations of theory into practice.

Hardcover ISBN: 978-1-53613-620-3
Retail Price: $195

Bone Regeneration: Concepts, Clinical Aspects and Future Directions

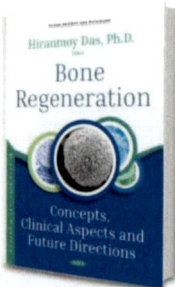

Editor: Hiranmoy Das

Series: Human Anatomy and Physiology

Book Description: This book provides an overview of the bone biology in normal homeostasis and in pathological conditions, along with clinical therapies currently considered and those being developed for future use.

Hardcover ISBN: 978-1-53613-990-7
Retail Price: $230